Daily Warm-Ups

MATH IN REAL LIFE

Brian Pressley

Level II

Certified Chain of Custody
Promoting Sustainable
Forest Management
www.sfiprogram.org

SGS-SFI/COC-US09/5501

The classroom teacher may reproduce materials in this book for classroom use only.
The reproduction of any part for an entire school or school system is strictly prohibited.
No part of this publication may be transmitted, stored, or recorded in any form
without written permission from the publisher.

1 2 3 4 5 6 7 8 9 10
ISBN 978-0-8251-6316-6
Copyright © 2007
J. Weston Walch, Publisher
40 Walch Drive • Portland, Maine 04103
www.walch.com
Printed in the United States of America

Table of Contents

Introduction . . . . *iv*

Consumer Math 1

Basic Math 57

Tables, Charts, and Graphs 106

Algebra and Geometry 120

Data Analysis and Probability 154

Answer Key 181

Daily Warm-Ups: Math in Real Life

The *Daily Warm-Ups* series is a wonderful way to turn extra classroom minutes into valuable learning time. The 180 quick activities—one for each day of the school year—practice applying a variety of math skills in real-life contexts. The warm-ups include an assortment of problems requiring varying degrees of prior knowledge—from Consumer Math to Algebra, Geometry, and Probability. The problems vary in the level of difficulty and the amount of time it will take students to solve them. Some of the problems are quite challenging, and students will need scientific calculators to solve some of them.

These daily activities may be used at the very beginning of class to get students into learning mode, near the end of class to make good educational use of that transitional time, in the middle of class to shift gears between lessons—or whenever else you have minutes that now go unused.

Daily Warm-Ups are easy-to-use reproducibles—simply photocopy the day's activity and distribute it. Or make a transparency of the activity and project it on the board. You may want to use the activities for extra-credit points or as a check on the math skills that are built and acquired over time. However you choose to use them, *Daily Warm-Ups* are a convenient and useful supplement to your regular lesson plans. Make every minute of your class time count!

Daily Warm-Ups: Math in Real Life

Adding Coins

Many transactions in real life involve the handling of money. The smallest units of money in the United States are the penny, the nickel, the dime, and the quarter. The penny is worth 1 cent, the nickel is worth 5 cents, the dime is worth 10 cents, and the quarter is worth 25 cents. 100 cents is equal to 1 dollar.

Determine the total value of each combination of coins below without using a calculator. Write the value on the line.

1. 1 nickel, 3 pennies, 2 dimes _____
2. 2 quarters, 2 dimes, 2 nickels _____
3. 5 nickels, 5 dimes, 3 pennies _____
4. 1 nickel, 1 penny, 1 quarter, 1 dime _____
5. 4 pennies, 12 nickels, 1 dime _____
6. 2 quarters, 2 pennies, 3 nickels _____

Daily Warm-Ups: Math in Real Life

Comparing Coins

Coins can be combined in many different ways. A large stack of coins may or may not have more value than a small stack of coins. For example, 50 pennies (50 cents) has less value than a much smaller stack of 11 nickels (55 cents). Sometimes a larger stack of coins does have a greater value than a small stack. For example, 75 pennies (75 cents) is worth more than 7 dimes (70 cents).

Determine which combination of coins below represents the greatest amount of money without using a calculator.

a. 53 pennies, 1 dime, 1 nickel

b. 1 quarter, 4 dimes, 1 nickel

c. 1 quarter, 3 dimes, 9 pennies

d. 2 quarters, 10 pennies, 1 nickel

e. 3 quarters

f. 5 dimes, 4 nickels, 3 pennies

Adding Bills

Amounts of money over 100 cents can be handled using bills. Common bills are the one-dollar bill, the five-dollar bill, the ten-dollar bill, the twenty-dollar bill, the fifty-dollar bill, and the hundred-dollar bill. A small stack of bills does not necessarily mean a small amount of money, and a large stack of bills does not necessarily mean a large amount of money.

Daily Warm-Ups: Math in Real Life

Determine the total value of each combination of bills below without using a calculator.

1. 1 five, 2 tens, 4 twenties _____
2. 2 hundreds, 3 fifties, 6 fives _____
3. 55 ones, 3 tens _____
4. 8 fives, 1 twenty, 6 ones, 8 hundreds _____
5. 12 fifties, 4 ones, 7 twenties, 3 fives _____
6. 3 ones, 1 five, 1 ten, 1 twenty, 1 fifty, 1 hundred _____

Comparing Bills

Bills can be combined in many different ways. A large stack of bills may or may not have more value than a small stack of bills. For example, 50 ones ($50) have less value than a much smaller stack of 11 fives ($55). Sometimes a larger stack of bills does have a greater value than a small stack. For example, 75 ones ($75) is worth more than 7 tens ($70).

Determine which combination of bills below represents the greatest amount of money without using a calculator.

a. 10 twenties, 10 fives, 20 ones

b. 40 fives, 1 ten, 20 ones

c. 2 hundreds, 1 fifty, 4 tens, 2 fives

d. 24 ones, 11 twenties

e. 4 fifties, 3 twenties, 18 ones

f. 100 ones, 2 fifties, 3 tens, 4 fives

Adding Coins and Bills

When you make a purchase, the total amount of money you spend is usually a combination of dollars and cents. An amount such as $10.45 or $8.55 is more common than a round number such as $11.00.

Determine the total value of each combination of bills and coins below without using a calculator.

1. 2 hundreds, 3 twenties, 1 ten, 7 nickels, 1 penny _____
2. 1 hundred, 8 twenties, 3 fives, 4 ones, 9 dimes, 8 pennies _____
3. 2 hundreds, 3 fifties, 52 ones, 1 quarter, 8 dimes, 1 nickel, 1 penny _____
4. 3 hundreds, 16 ones, 8 nickels _____
5. 4 fifties, 6 twenties, 4 tens, 1 five, 4 ones, 9 quarters _____
6. 6 twenties, 3 fives, 10 ones, 2 dimes, 12 nickels _____

Comparing Coins and Bills

The value of an amount of money is not always related to the size of the stack. For instance, 5 hundreds are worth as much as a stack of 50,000 pennies. The 5 hundred-dollar bills would weigh about 0.01 pounds, and the 50,000 pennies would weigh about 275 pounds! The values are the same even though the pennies weigh 27,500 times more than the bills and take up a lot more room.

Determine which combination of bills and coins below represents the greatest amount of money without using a calculator.

a. 10 twenties, 3 fives, 10 ones, 2 dimes, 2 pennies
b. 1 hundred, 4 twenties, 3 fives, 4 ones, 9 dimes, 1 penny
c. 2 hundreds, 3 fifties, 12 ones, 1 quarter, 1 dime, 1 nickel, 1 penny
d. 2 hundreds, 3 twenties, 1 ten, 3 nickels, 1 penny
e. 2 hundreds, 16 ones, 8 nickels
f. 4 fifties, 4 twenties, 1 ten, 1 five, 4 ones, 4 quarters

© 2007 Walch Publishing

Daily Warm-Ups: Math in Real Life

Making Change with Coins

Often when you buy something with coins, you will not have the exact change. Therefore, you will give more than you owe and get change back. It is a good idea to figure out how much money you are owed to make sure you are getting the correct change.

Determine the amount of change you should get back for each of the following purchases without using a calculator.

1. You buy 3 pencils at 15¢ each and pay with 2 quarters. _____

2. You pay a 60¢ toll with 75¢. _____

3. You buy a pack of gum for 59¢ and pay with 12 nickels. _____

4. You buy a bottle of water for $1.29 and pay with 5 quarters and 1 dime. _____

5. You buy a bottle of apple juice for 99¢ and pay with 3 quarters and 3 dimes. _____

Making Change with Coins and Bills

Often when you buy something with cash, you will not have the exact change. Therefore, you will give more than you owe and get change back. It is a good idea to figure out how much money you are owed to make sure you are getting the correct change.

For each problem below, determine whether or not you received the correct change without using a calculator.

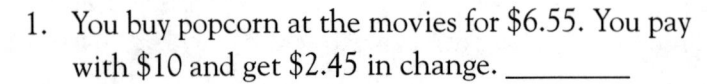

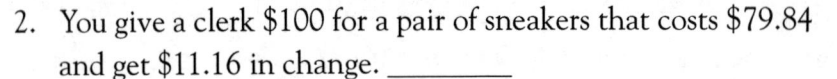

1. You buy popcorn at the movies for $6.55. You pay with $10 and get $2.45 in change. _____

2. You give a clerk $100 for a pair of sneakers that costs $79.84 and get $11.16 in change. _____

3. You spend $16.52 on magazines and pay with 2 tens. You get $3.48 in change. _____

4. You buy a meal at a fast-food restaurant for $7.58. You pay with 1 twenty and get $13.42 in change. _____

5. You buy some DVDs at a flea market for $64.91. You give 4 twenties and get $15.09 in change. _____

Daily Warm-Ups: Math in Real Life

Multiplying Money

Sometimes when an item is being sold at a reduced price, it is a good deal to buy many of that item. To find the total cost of your purchase, you can multiply the cost of one item by the number of items you're buying.

For each problem below, determine the total cost.

1. You buy 6 sodas for $.79 each. _____
2. You buy 14 sub sandwiches for $2.69 each. _____
3. You find basketball sneakers on sale for $42.35 a pair, so you buy 3 pairs. _____
4. You buy 30 rolls of toilet paper for $0.52 a roll. _____
5. You're making a lot of cookies and buy 12 bags of chocolate chips, which are $1.89 per bag. _____
6. You want to download 34 songs for $0.99 each. _____

Dividing Money

Sometimes when you buy items in bulk, you save money. But are you always getting the best deal? The best way to know is to calculate the cost per item. For example, if you buy a six-pack of soda for $1.92, then each can costs 32¢. If you purchase the cans from a vending machine, the cost could be a dollar or more per can. Buying the six-pack of soda is the better deal.

Daily Warm-Ups: Math in Real Life

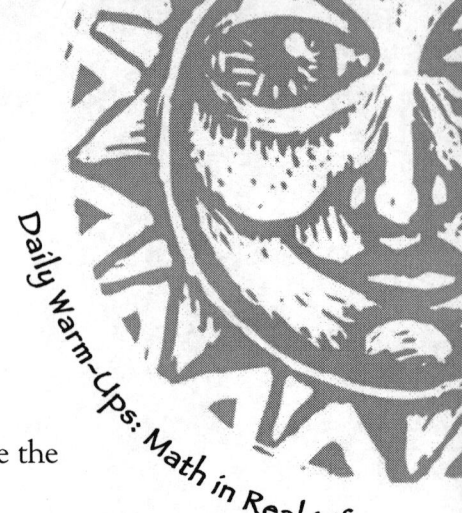

For each item below, determine which choice is the better buy. Circle the correct choice.

1. a six-pack of paper towels for $6.50 or six rolls for $1.25 each
2. a 24-can case of soup for $18 or a 4-can pack of soup for $3
3. a 12-pack of pencils for $0.99 or a 50-pack for $4.50
4. a 500-sheet ream of paper for $5.99 or a 250-sheet ream for $3.50
5. a 50-pack of DVD-RWs for $40 or a 12-pack for $10.99
6. a whole album download (14 songs) for 9.99 or a single track for 75¢ each

10

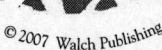

Daily Warm-Ups: Math in Real Life

Estimating a Bill

When you go to a restaurant with a group of people, you often have to determine who owes what when the bill arrives. Sometimes the bill is split up for you. If it is not, however, how do you determine your fair share? One of the simplest ways is to round each item you bought to the nearest whole dollar and then add them together.

Estimate each person's share of the bill below. Each person has a different seat number next to his or her items. After you make your estimates, find the exact amounts owed.

Seat 1: chicken burger $6.95
Seat 2: BLT $4.95
Seat 3: salad special $5.29
Seat 3: bottled water $1.49
Seat 1: soda $1.75
Seat 2: iced tea $1.55
Seat 1: chicken wings $3.59
Seat 3: coconut shrimp $4.59
Seat 2: slice apple pie $2.59
Seat 3: ice cream sundae $3.09

Discounts

A discount is a way for a business to offer you a service or a product at a lower cost than normal. Discounts are often found when a store is having a sale or trying to move items that have not been selling well.

Determine the discount amount and the sale price for each of the items below.

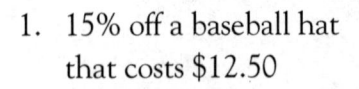

1. 15% off a baseball hat that costs $12.50

    discount = _____

    sale price = _____

2. 20% off a portable stereo that costs $45.95

    discount = _____

    sale price = _____

3. 50% off a gold necklace that costs $375.00

    discount = _____

    sale price = _____

4. 15% off a new television that costs $2900.00

    discount = _____

    sale price = _____

Unit Pricing: Lengths

There are a number of items that you can buy by length. For example, there are places where necklaces, rope, and fabric are sold by the inch, foot, or yard. To determine the total cost, you multiply the cost per length by the total length (such as the number of dollars per foot by the total number of feet).

Find the total cost of each of the following items that are being sold by length.

1. 250 feet of rope at $0.35 per foot _____
2. 18 inches of gold necklace at $35 per inch _____
3. 6 yards of denim fabric at $3.79 per yard _____
4. 160 feet of lumber at $2.49 per foot _____
5. 800 feet of coaxial cable at $0.42 per foot _____
6. 60 feet of copper pipe at $1.25 per foot _____

Unit Pricing: Area

There are a number of items that you can buy by area. For example, there are places where carpets, rugs, and even sod are sold by the square foot or square yard (equal to 9 square feet). To determine the total cost, you can multiply the cost per unit of area by the number of units (such as the number of dollars per square foot by the number of square feet).

Daily Warm-Ups: Math in Real Life

Find the total cost of each of the following items that are being sold by area.

1. 100 square feet of canvas at $1.19 per square foot _____
2. 800 square feet of carpeting at $2.29 per square foot _____
3. 25 square feet of poster board at $0.79 per square foot _____
4. 6 square yards of fiberglass cloth at $13.99 per square yard _____
5. 4 square yards of fabric for a dress at $6.65 per square yard _____
6. 2100 square feet of red oak hardwood flooring at $4.49 per square foot _____

© 2007 Walch Publishing

Unit Pricing: Volume

There are a number of items that you can buy by volume. For example, there are places where gasoline, milk, and even air are sold by the liter, gallon, cubic foot, or yard. In this case, a yard is a cubic yard, 3 feet high, 3 feet wide, and 3 feet deep, or 27 cubic feet. To determine the total cost, you can multiply the cost per unit of volume by the number of units (such as the number of dollars per cubic foot by the number of cubic feet).

Daily Warm-Ups: Math in Real Life

Find the total cost of each of the following items.

1. 3 gallons of milk at $3.69 per gallon _____

2. 45 liters of helium at $0.42 per liter _____

3. 3 yards of gravel at $125 per yard _____

4. 6 liters of soda at $1.39 per liter _____

5. 24 quarts of oil at $1.79 per quart _____

6. 100 gallons of water at $1.19 per gallon _____

Unit Pricing: Weights

There are a number of items that you can buy by weight. For example, there are places where food, nails, or scrap metal are sold by the ounce, pound, or ton. To determine the total cost, you can multiply the cost per unit of weight by the number of units (such as the number of dollars per ounce by the number of ounces).

Find the total cost of each of the following items that are being sold by weight.

1. 12 pounds of nails at $2.25 per pound _____
2. 2.2 pounds of chicken at $4.49 per pound _____
3. 4 ounces of gold chain at $650 per ounce _____
4. 4 tons of gravel at $180 per ton _____
5. 25 pounds of dog food at $1.29 per pound _____
6. 10 one-ounce silver coins at $29.95 per ounce _____

Deals by Length

Not every store gives the price per length for the various items they are selling. For example, which is the better deal: 100 feet of rope for $16.99 or 250 feet of rope for $40? The best way to find out is to calculate the unit price per length. You can do this by dividing the total cost by the length of the item.

Determine which of the two choices for each item below is the better buy. Circle the correct choice.

1. 150 feet of rope for $18.50 or 250 feet of rope for $31.50
2. an 18-inch gold chain for $625 or a 22-inch gold chain for $775
3. 6 yards of cotton fabric for $21.24 or 8 yards of cotton fabric for $25.39
4. a 100-foot coil of chain for $55.20 or a 175-foot coil of chain for $85.40
5. ten 8-foot boards for $35.89 or fifteen 8-foot boards for $58.21
6. 75 feet of PVC pipe for $111.50 or 100 feet of PVC pipe for $135.25

Deals by Area

If you know that you need 500 square feet of fabric to make blankets for a charity, you would like to get the best price. If you buy fabric in bulk, you may not be given the price per square foot. To find out if you are getting a good price, you can divide the total cost by the number of square feet to find out the cost per square foot of fabric.

Determine which of the two choices for each item below is the better buy. Circle the correct choice.

1. 100 square feet of denim for $18.50 or 150 square feet of denim for $26.75
2. 1100 square feet of sod for $1200 or 1000 square feet of sod for $1100
3. 250 square feet of hardwood flooring for $1425 or 265 square feet of flooring for $1620
4. 820 square feet of carpet for $1222 or 850 square feet of carpet for $1300
5. paint to cover 400 square feet for $25.65 or paint to cover 450 square feet for $31.45
6. a 24-square-foot table for $655 or a 32-square-foot table for $820

Daily Warm-Ups: Math in Real Life

Deals by Volume

The grocery store is filled with items that are sold by gallons, half gallons, quarts, pints, and fluid ounces. Many stores give the cost per unit on the shelf near the item, but often items on sale or discounted items are missing such information.

Determine which of the two choices for each item below is the better buy. Circle the correct choice.

1. 32 ounces of maple syrup for $3.99 or 48 ounces of syrup for $4.99
2. a gallon of milk for $3.98 or a half gallon of milk for $1.99
3. a 16-ounce can of carrots for $1.29 or a 20-ounce can of carrots for $1.49
4. 3 quarts of apple juice for $3.59 or 1 gallon (4 quarts) of apple juice for $3.99
5. a half gallon of pineapple juice for $1.79 or 1 gallon of juice for $3.55
6. 64 ounces of spaghetti sauce for $4.60 or 48 ounces of sauce for $3.79

Deals by Weight

In a grocery store, many items are sold by weight at the deli counter. But in many grocery stores, these same items are sold prepackaged at a better price. Just as ham or turkey is sold by the pound at the deli counter, it is also often sold in bulk packages in coolers elsewhere in the same store.

Determine which of the two choices for each item below is the better buy. Circle the correct choice.

1. $4.29 per pound for ham or 2 pounds of ham for $8.50
2. $3.50 per pound for potato salad or 1.5 pounds of salad for $4.50
3. $2.29 per pound for mozzarella cheese or $\frac{1}{2}$ pound of cheese for $1.29
4. $3.79 per pound for chicken wings or $20.25 for a 5-pound bag
5. $6.89 per pound for salmon or $3.50 for 8 ounces (half a pound) of salmon
6. $14.59 per pound for lobster or 3 pounds of lobster for $40.99

Living Expenses

When you are on your own and depending on your own income, you may find that the money disappears almost as fast as you can make it. Many people find it useful to make a budget to see where their money is going.

In the space below, make a list of all the expenses you might expect to incur when you are on your own. Some examples of possible expenses are property taxes, life insurance, mortgage, and car payments. Once you have made a list, group the expenses into categories such as entertainment, clothing, food, rent, and so forth.

Paychecks

When you receive your first real paycheck, you may be disappointed by the amount of money that you **don't** get. The gross amount of your paycheck is the total you are paid, while the net amount is the amount you actually "take home." Amounts of money are withheld for things such as state and federal taxes, Social Security, health or life insurance, a retirement plan, Medicare, union dues, and other costs. These withholdings are sometimes specific to the job you are getting paid for.

Determine what percent of each person's check is withheld by dividing his or her net pay by the gross pay.

1. Shelly: gross pay $1254.25, net pay $901.22 _____
2. Karyn: gross pay $650.01, net pay $414.59 _____
3. Maku: gross pay $1854.26, net pay $1225.90 _____
4. Sergio: gross pay $1187.25, net pay $785.25 _____
5. Horatio: gross pay $359.68, net pay $259.24 _____
6. Jayden: gross pay $777.85, net pay $489.25 _____

Daily Warm-Ups: Math in Real Life

Loans and Simple Interest

When you have a savings account, the bank pays you a small amount, called interest, for having the use of your money. The bank loans this money at a higher interest rate, and that money can be used by people to buy cars, houses, and so forth. A loan with simple interest is one in which the borrower agrees to pay a certain percent of the amount borrowed. So a $600 loan for one year at 7% would be paid back as $600 plus 7% of $600, or $42. The interest is $42, and the total amount due is $642.

Calculate the simple interest on each of the following loans.

1. $35,000 at 1.5% _____

2. $950 at 11% _____

3. $125,000 at 5.75% _____

4. $12,675 at 18% _____

5. $5599 at 3.5% _____

6. $8500 at 18.99% _____

Loans and Compound Interest

Compound interest applied to a loan or a savings account is interest that is paid on the original amount of money and on the interest that has built up. The formula to find the total amount of money after all of the interest has been applied is $T = P(1 + r)^n$. In this formula, T is the total amount of money, P is the initial money or principal, r is the interest rate expressed as a decimal, and n is the number of times the interest is applied. For yearly interest, for example, n would be the number of years.

Daily Warm-Ups: Math in Real Life

Determine the total amount of money after the compound interest has been applied for each situation below.

1. $12,000 at 3% for 2 years _____

2. $500 at 1.79% for 9 years _____

3. $50 at 2% for 50 years _____

4. $31,588 at 5.5% for 3 years _____

5. $8210 at 12% for 6 years _____

6. $900.50 at 6.69% for 11 years _____

Credit Cards

Credit cards are well known for having relatively high interest rates. Credit card companies will often offer a period of time in which you pay little or no interest to get you to use their card. When this grace period runs out, however, you are left with a much higher interest rate. To figure out your monthly interest rate, take your yearly interest rate and divide it by 12. This is the rate that will be applied to your balance each month.

Daily Warm-Ups: Math in Real Life

Determine the amount of interest that would be paid on each of the following credit card accounts for one month.

1. $355.69 at 18.99% yearly interest _____
2. $2000.22 at 21.5% yearly interest _____
3. $1243.44 at 17.9% yearly interest _____
4. $987.50 at 16.69% yearly interest _____
5. $39.92 at 20.75% yearly interest _____
6. $90.91 at 14.00% yearly interest _____

© 2007 Walch Publishing

Term Loans

A term loan is a loan in which the initial principal and the interest are all paid back in one payment after a fixed amount of time. Term loans are often for a one-year period.

Determine the total amount that will have to be paid back for each of the term loans below.

1. $13,200 at 3.5% _____
2. $700 at 1.9% _____
3. $750 at 2.6% _____
4. $51,522 at 4.5% _____
5. $8710 at 1.229% _____
6. $9100 at 8.89% _____

Interest-Only Loans

An interest-only loan is a loan in which a portion of the interest is paid each month. Then the principal is paid back in a lump sum at the end of the term of the loan. This is a way of borrowing money when you know that you are going to be able to pay off a larger lump sum all at once but will only be able to make small payments in the near future. To determine your monthly payment, you calculate the total yearly interest on the loan and then divide by the number of months you will be paying interest.

Determine the monthly interest payment and total payment based on the number of years for each of the following interest-only loans.

1. principal = $12,500   interest = 5%   per year term = 3 years

   monthly interest payment = _____

   total payment = _____

2. principal = $3200   interest = 6%   per year term = 2 years

   monthly interest payment = _____

   total payment = _____

Personal Checks

Personal checks can be used to pay bills through the mail or in situations in which it is not convenient to use cash. A personal check has a lot of information on it.

```
Jon Dough                      Date 6/12/09         0293
400 Carlson Street                                  22-222
Brunswick, ME 04011                                  560
555-555-1234

PAY TO THE   A-1 Motor Services          $ $421.65
ORDER OF
Four hundred twenty-one 65/100                   DOLLARS

       North Bank
       Portland, ME
FOR  car repairs                    Jon Dough

⑆416283841⑆  007227327⑈  0293
```

Answer the following questions about the personal check above.

1. Whose bank account is this check from? _____
2. Who is the check made out to? _____
3. What amount is the check made out for? _____
4. What was the check for? _____
5. At what bank is this account found? _____
6. What is the account number on this check? _____

© 2007 Walch Publishing

Check Register

In a check register, you keep track of the checks you write that take money out of your checking account. These are called debits. You can also add in interest earned on the account and money deposited from paychecks or gifts. These are called credits.

DATE	CHECK NO.	DESCRIPTION OF TRANSACTION	AMOUNT OF DEBIT		AMOUNT OF CREDIT		BALANCE 621	42
12/6	1241	Music Superstore	24	11				
12/6	1242	Tanya's Hair Salon	34	50				
12/7	1243	Sweet Dreams Candy	6	21				
12/7		Paycheck			485	21		
12/7	1244	Smith's Car Repair	372	11				
12/10		Gift from Auntie			25	00		

Figure out the balance in the check register above. Then answer the questions that follow.

1. How much money is in the account after the gift from Auntie is added?
2. When was the check written for car repair?
3. What check number was used at the Music Superstore?
4. How much money was in the account before the trip to the Music Superstore?

Savings Accounts

A savings account differs from a checking account. Generally, a savings account is used to keep money you do not need to spend in the near future. Interest on a savings account is usually higher than on a checking account. Many banks do not offer interest on checking accounts.

For each of the following, determine how much more interest you would earn in a year if your money was in a savings account instead of in a checking account.

1. $1000 in savings at 4% interest or in checking at 0.5% interest _____
2. $12,678 in savings at 2.25% interest or in checking at no interest _____
3. $500 in savings at 3.29% interest or in checking at 0.65% interest _____
4. $21,454 in savings at 2.79% interest or in checking at 0.49% interest _____
5. $255,683 in savings at 2.3% interest or in checking at 1.1% interest _____

Tax Percents

You are required to pay federal income tax on the money you earn at any job. Below is a chart that shows the marginal tax rate for federal income tax for the year. A marginal tax rate is one that goes up along with the amount of income. As your income gets higher, the amount of taxes you pay increases as well.

If your filing status is **Single**

Taxable Income Over ---	But not over---	Marginal Rate
$0	$7,300	10%
$7,300	$29,700	15%
$29,700	$71,950	25%
$71,950	$150,150	28%
$150,150	$326,450	33%
$326,450	and over	35%

Answer the following questions about the chart above.

1. What was the highest tax rate as a percent?

2. How much would you have paid in federal income tax if you had made $89,565?

3. What is the tax rate on $151,151?

4. Is the tax on $50,000 double the tax on $25,000? Why or why not?

Inflation

Over time, the value of a dollar changes. When your grandparents were teenagers, a soda cost a nickel. That sounds like a good deal. But back then the average salary in the United States was $1299 a year. In 2005, the average salary was $40,409 a year. Below is a chart showing what a 2006 dollar would be worth as far back as 1990.

Daily Warm-Ups: Math in Real Life

Year	Value of a dollar relative to 2006	Year	Value of a dollar relative to 2006
1990	0.649	1999	0.827
1991	0.676	2000	0.855
1992	0.696	2001	0.879
1993	0.717	2002	0.893
1994	0.735	2003	0.913
1995	0.756	2004	0.937
1996	0.779	2005	0.969
1997	0.797	2006	1.000
1998	0.809		

Use the chart to complete the following problems.

1. How much does the 2006 dollar lose in value each year back to 1990?

2. Predict the value of a dollar each year from 2007 to 2010.

3. How much value did the dollar lose from 1990 to 1998? From 1990 to 2006?

© 2007 Walch Publishing

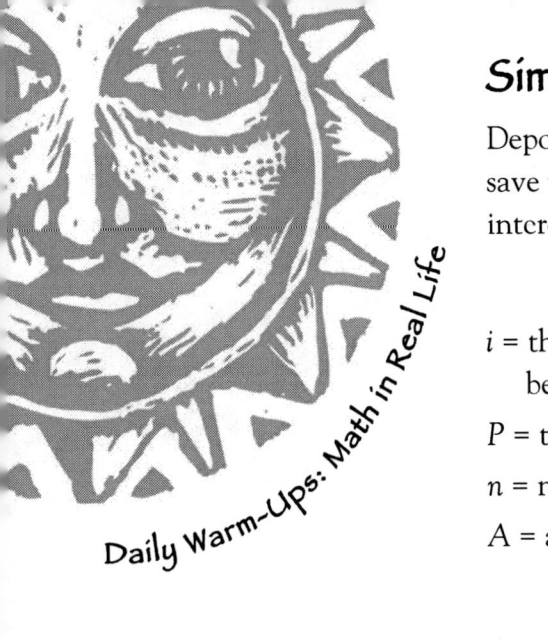

Simple Interest

Depositing money in an interest-bearing account is a good way to save for college. Use the formula below to determine how much interest each student will earn.

$$A = P(ni)$$

i = the yearly interest rate (as a decimal, e.g., a rate of 6% would be 0.06)

P = the amount of the principle (original deposit)

n = number of years

A = amount after n years

1. $3000 at 4% interest for 4 years

2. $5000 at 5% interest for 12 years

3. $10,000 at 6% interest for 2 years

Car Insurance

You cannot drive off the lot at a car dealership until you can show proof of insurance. There are a number of insurance companies competing for your business on television, in the newspaper, on the Internet, in the mail, and on the radio. Certain factors about your life determine what your insurance rate will be. For instance, do you live in the city or in the country? Do you have an expensive car with lots of safety features, or do you drive a run-down old jalopy?

In the space below, list all the factors that might require you to pay more (or less) for your car insurance. There are at least seven factors that most insurance companies use.

Home Insurance

Owning a home is a costly experience. Few people can afford to buy a house without taking out a loan. Even fewer can afford to replace a house that burns down with all their possessions inside of it. Homeowner's insurance allows you to replace your home and even its contents if they are destroyed or damaged in a way that is described in your insurance policy. A typical home is insured for 80% of its value, and the contents are assumed to be worth 50% of the house value.

Answer the following questions.

1. Your house is worth $275,000. How much should it be insured for? How much should the contents be insured for?

2. Your house is worth $185,000. It's insured for $150,000. Is that enough?

3. The contents of your $328,000 house are insured for $164,000. Is that enough?

4. Your home is worth $175,000. What should the house and its contents be insured for?

Life Insurance

Life insurance pays money to a person's dependents when he or she dies. There are many potential costs to a family after the death of a loved one. These include funeral costs, possible medical costs from an expensive hospital stay, and day-to-day expenses such as food, clothing, and housing if the person who died was a major source of income. The general rule is to have six times your gross income in life insurance.

In each of the following cases, determine if the person has enough life insurance. If the person does not have enough, determine how much more he or she needs.

1. gross income = $45,621, life insurance = $275,000

2. gross income = $155,350, life insurance = $1,000,000

3. gross income = $21,225, life insurance = $100,000

4. gross income = $35,261, life insurance = $175,000

5. gross income = $95,450, life insurance = $575,000

6. gross income = $54,540, life insurance = $300,000

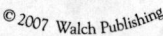

Comparing Business Rates

You can save a lot of money by shopping around, not just for items such as cars, but also for services that you receive from businesses. For example, say you need someone to paint your house. If one company charges by the square foot and the other by the hour, there could be a significant difference in the prices they might charge you.

In each case below, determine the best buy for the service described. Circle the best choice.

1. One carpenter will put in your 300 square feet of hardwood flooring for $2.50 per square foot, while the other guarantees to install it in twelve hours, at $70 per hour.

2. Technical support from your computer company for 3 years is $59.95, while a rival company offers 2 years of service at $49.95.

3. A cell phone company is running a special that offers 1000 minutes a month for $19.95, or you could buy a calling card that has 750 minutes for $12.50.

Monetary Conversions

Traveling to a different country can be an eye-opening experience. You may encounter different foods, music, art, books, movies, and social customs. Sometimes the hardest thing to get used to is the money. Exchange rates exist that allow you to change your money into the money of the country you are visiting. Sometimes the exchange rate works in your favor, and sometimes you actually lose value when you change to the new money.

Complete each of the monetary conversions below using the currency conversion chart.

Currency	Yen	Euro	Can $	U.K. £	AU $	Swiss Franc
1 U.S. $ =	116.2900	0.7877	1.1364	0.5394	1.3139	1.2421

1. $4.96 to yen
2. $300 to euros
3. $155.68 to Canadian dollars (Can $)
4. $12,560 to pounds (U.K. £)
5. $850 to Australian dollars (AU $)
6. $1500 to Swiss francs

Taxes

Everyone who earns money pays taxes. For those earning less than $25,000 per year, the rates are as follows:

15% Federal Income Tax
9% FICA and Medicare
3% State Income Tax

Determine the amount of taxes paid on one week's paycheck for each employee below.

1. working in a pet store for 20 hours per week at $7.50 per hour

2. washing dishes in a restaurant for 15 hours per week at $6.25 per hour

3. teaching swimming for 8 hours per week at $10.00 per hour

Comparison Shopping

You can save a lot of money by comparison shopping. For small items such as candy bars or pencils, you might not see a large difference in price when you comparison shop, but even these small savings can add up over time. For larger purchases, it's often immediately clear that you are saving a significant amount. Comparing sales or checking for deals on the Internet can be beneficial to your bank account.

Determine which store has the best deal for each of the following purchases.

Item	Store 1	Store 2
1. video camera	$499	$475 online + $29.95 shipping
2. sweatshirt	$38.50	$42 on sale for 15% off
3. MP3 player	$99.99	$79.99 online + $7.95 shipping
4. bookcase	$108	2 for $235
5. CD	$12.99	$14.50 on sale for 30% off
6. cell phone	$59.95	$59.95 + free in-car charger

Paycheck Deductions

Your paycheck has a number of deductions to pay for things such as taxes and retirement costs. Net pay, or take-home pay, is pay after taxes and other amounts have been taken out.

Determine how much each person below will get to take home after his or her deductions are made.

1. Malcolm makes $598.50 each week. He pays 11.5% in federal income tax, 5.3% in state income tax, and 7.65% for Social Security and Medicare. What is his net pay?

2. Kendie makes $695.70 each week. She pays 14% in federal income tax, 11% into a retirement fund, and 1.8% in city taxes. What is her take-home pay?

3. Jack makes $1370.00 each week. He pays 14.2% in federal income tax and has $150 taken to pay for a legal settlement. What is his net pay?

4. Lila makes $322.60 each week. She pays 13.5% in federal income tax, 3.5% in state income tax, and $50 toward her retirement. What is her take-home pay?

Pay Rates

Entry-level jobs do not pay extremely well. As people get more experience and education, they often find that higher-paying jobs are open to them. Some people get paid an hourly rate, while others are paid a salary that is a flat fee, usually for a year of work at a time. If you get an hourly rate, you can find out how much you are getting at the end of a pay period by multiplying the number of hours you've worked by the hourly pay rate you receive.

Solve the following pay rate problems.

1. How much will you get paid if you work 28.5 hours for $7.75 per hour?

2. Your salary is $65,788. If you work fifty 40-hour weeks in a year, what is your hourly wage?

3. You work 4 hours on Monday, 4.5 hours on Tuesday, 7.25 hours on Thursday, and 12 hours on Saturday. If you get paid $11.52 per hour, what is your total pay for the week?

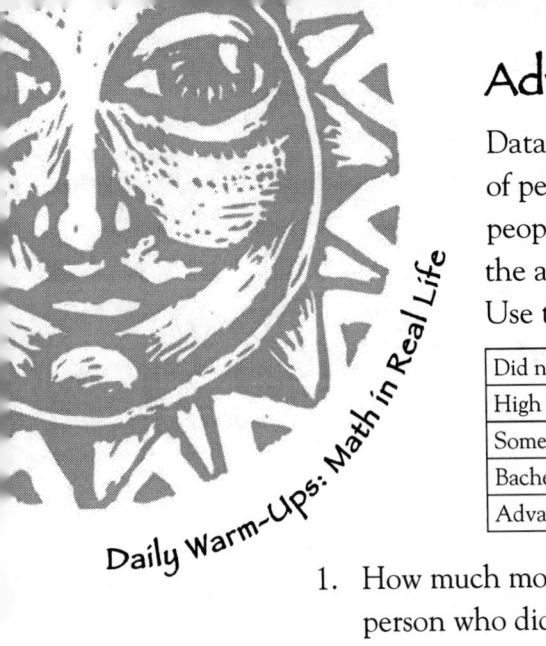

Advantage: Education

Data compiled by the U.S. Census Bureau show that the majority of people who have advanced degrees make more money than people who did not finish high school or even college. Below are the average yearly salaries from one state based on education level. Use the chart to answer the questions that follow.

Did not graduate high school	$22,836
High school graduate	$29,586
Some college	$34,654
Bachelor's degree	$46,776
Advanced degree	$58,477

1. How much more does a person with a bachelor's degree earn than a person who did not graduate from high school?

2. What percent more does a person with a bachelor's degree earn than a person with only some college?

3. How many times more money does someone with an advanced degree earn than someone who only graduated from high school?

4. By what percent does income increase between each of the five levels listed above?

Comparing Phone Plans

There are so many cell phone plans that it is almost impossible to compare them all. Not only do competing companies have different plans, but many companies offer multiple plans. One way to determine the best plan for you is to imagine a worst-case scenario in which you use your phone far more than you ever think you might. You should consider text messaging charges, Internet charges, and any other service offered by your phone company that you will be charged for.

Use the chart below to calculate how much each plan would cost in the unlikely event that you used 6000 minutes in one month. (That's 4 days and 4 hours!)

Plan	Monthly cost	Minutes	Extra minutes	Total cost
1	$29.99	500	45¢ each	
2	$49.99	1000	40¢ each	
3	$64.99	1500	35¢ each	
4	$74.99	2000	30¢ each	
5	$129.99	4000	20¢ each	

Mortgage Rates

One of the largest purchases you will probably ever make is a home. The purchase of a home usually requires that you get a mortgage. A mortgage is an agreement you make with a bank in which they loan you money to buy a house. If you fail to make payments on time, the house will become the property of the bank. Be sure you can afford the monthly payments on the home you wish to purchase.

Complete the following.

1. Find an online mortgage calculator and determine the monthly payment for a $350,000 home at 6%, 6.5%, 7%, 7.5%, and 8% interest if the loan is for 30 years.

2. Multiply the monthly payment by 360 to find the total amount paid over the life of the loan for each loan percentage.

3. How much more will it cost over the life of the loan if you have a 7% interest rate instead of a 6% interest rate?

4. How much more will it cost over the life of the loan if you have an 8% interest rate instead of a 6% interest rate?

Recipes

The amounts given in recipes are often in cups, tablespoons, teaspoons, sticks of butter, and number of eggs. Below is a recipe for coconut macaroons.

$\frac{2}{3}$ cup sweetened condensed milk $\quad$ $2\frac{2}{3}$ cups shredded coconut

$\frac{1}{8}$ teaspoon salt $\quad\quad\quad\quad\quad\quad\quad\quad$ 1 teaspoon vanilla extract

Drop by the spoonful on a baking sheet and cook at 350° until golden brown.

Answer the following questions.

1. How much of each ingredient would you need if you doubled the recipe?
2. How much of each ingredient would you need if you halved the recipe?
3. How much coconut would you need for 5 batches of macaroons?
4. How much salt would you need for 8 batches of macaroons?
5. How much milk would you need for 4 batches of macaroons?
6. How much coconut would you need for 24 batches of macaroons?

Commission

Working on commission is a way that you can earn money that goes beyond hourly wage or simple salary. When you are working in sales, your income may be completely or partially determined by the amount you sell or the value of the goods or services you sell. In some circumstances, you might earn an hourly wage that is relatively small and a commission on everything you sell.

Suppose you work at a car dealership where they pay you $10.50 per hour with no special overtime rate. Each week, you get 5% of your first $80,000 in sales, 7% of your second $80,000 in sales, and 10% of any sales over $160,000. Determine the weekly pay you would receive in each of the following circumstances.

1. 20 hours of work and $52,000 in sales _____

2. 60 hours of work and $221,000 in sales _____

3. 25 hours of work and $75,000 in sales _____

4. 18 hours of work and $138,000 in sales _____

5. 30 hours of work and $352,000 in sales _____

6. 84 hours of work and $452,000 in sales _____

To Insure Prompt Service

It is considered polite to tip your waitperson after he or she has waited on you in a restaurant. Tips are often 10%, 15%, or 20% of the bill, or even more. If calculating a tip in your head is a challenge, remember that if you take the bill's total and move the decimal one place to the left, you are looking at 10% of the bill. Then you can double that amount to get 20% of the bill. The amount halfway between these two amounts is 15% of the bill.

Determine the tip amount on each of the following bills. Try to estimate the tip without a calculator, and then check your estimate afterward.

1. 10% for a meal that costs $43.50 _____
2. 15% for a meal that costs $21.95 _____
3. 20% for a meal that costs $65.89 _____
4. 10% for a meal that costs $32.32 _____
5. 15% for a meal that costs $125.85 _____
6. 20% for a meal that costs $12.69 _____

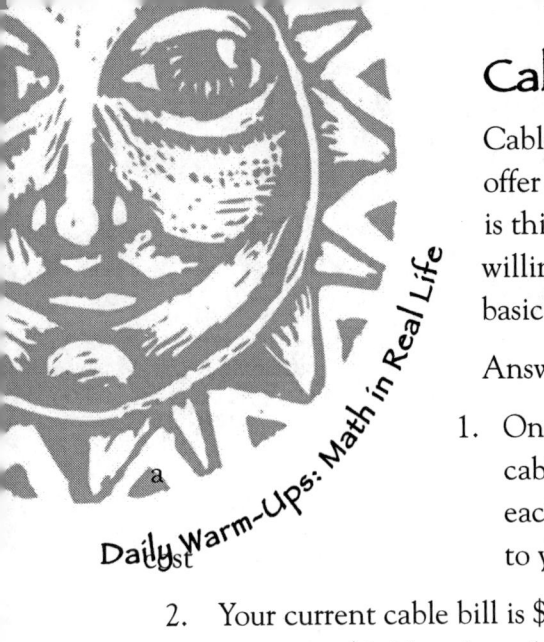

Cable Concerns

Cable companies are in direct competition with one another. They offer many services and channels, but the question that matters is this: Do they offer the channels you want for the price you're willing to pay? Companies offer pay-per-view events and movies, basic packages, cable box rental, premium channels, and more.

Answer the following questions.

1. One cable company offers a basic package of channels for $29.95, cable box for $3.00, two premium channels you want for $9.99 each, and a remote for the cable box for $1.50. What is the total cost to you on your first bill?

2. Your current cable bill is $38.50 per month. You order 4 pay-per-view movies at $3.95 each, order a live pay-per-view concert for $20.95, and add a premium channel for $11.50 per month. What will your next cable bill be?

3. Your last cable bill was $45.90. You decide to get cable Internet service through your cable provider. The installation is free, but the service is $29.95 per month, and renting a cable modem costs $2.95 per month. What will your next cable bill be?

Big-Screen TV

You are interested in buying a big-screen television. You want the largest amount of screen for your money. You are considering different options, including plasma televisions, rear-projection televisions, LCD televisions, and tube televisions. Below are prices for some different televisions. Their size measurements are given as the diagonal distance across the screen.

55-inch plasma television: $4299.99

46-inch LCD television: $3299.99

61-inch rear-projection television: $2699.99

42-inch tube television: $899.99

Answer the following questions.

1. Which television has the best price per diagonal inch?

2. Which television has the worst price per diagonal inch?

3. Why might diagonal distance not be the best way to determine whether or not you're getting the best deal?

Air Conditioning and BTUs

BTU is short for British thermal unit and is a way of measuring heat. It represents the amount of heat energy needed to raise the temperature of 1 pound of water 1° Fahrenheit. The more BTUs an air conditioner is rated for, the larger amount of space it is capable of cooling.

Below is a chart of the BTU ratings for nine air conditioners. Calculate the number of BTUs per dollar for each one. Then determine which air conditioner has the highest number of BTUs for each dollar you spend.

Air Conditioner	BTUs	Price	BTUs per dollar
1	24,000	$388	
2	18,000	$324	
3	17,000	$488	
4	15,000	$398	
5	12,000	$238	
6	10,000	$194	
7	8,000	$157	
8	6,000	$128	
9	5,000	$98	

Refrigerator Size

Refrigerators come with a large variety of features, from ice makers to built-in wireless LCD screens that can receive images from the Internet or your television. The arrangement of the inside of the refrigerator has a lot to do with how much you can put into it. Refrigerators have their available space measured in cubic feet. If you had a box that had 1 cubic foot of space in it, it would hold almost 7.5 gallons of water! A cubic foot might not sound like much, but it's a lot of room in a refrigerator.

Determine which refrigerator below has the most cubic feet per dollar.

a. 25.1 cubic feet for $1149.99

b. 18.6 cubic feet for $749.99

c. 22 cubic feet for $1049.99

d. 22.4 cubic feet for $1149.99

e. 22.1 cubic feet for $949.99

f. 19.5 cubic feet for $979.99

Medicine Dosage

When adults take medicine, they usually take a number of tablespoons, teaspoons, or pills every few hours. However, young children cannot have adult doses of medicine when they are sick. Doctors have to prescribe pediatric doses based on the mass of the child in kilograms. To find the mass of a child in kilograms, you divide the child's weight in pounds by 2.2.

Determine the correct dosage of medicine for each child below.

1. 40-pound boy   dosage: 12 milligrams/kilogram
2. 16-pound girl   dosage: 15 milligrams/kilogram
3. 52-pound girl   dosage: 50 milligrams/kilogram
4. 70-pound boy   dosage: 8 milligrams/kilogram
5. 9-pound girl   dosage: 45 milligrams/kilogram
6. 22-pound boy   dosage: 30 milligrams/kilogram

Misleading Advertising

You hear claims made about products on television and on the radio, and see similar claims in newspapers, magazines, and on the Internet. If you look carefully at the claims that are made, you might be able to see how an advertising agency or company could manipulate facts or numbers to persuade you to buy their product.

For each of the following claims, determine how the advertiser might have been manipulating data to sell you their product.

1. Four out of five doctors recommend our crutches.
2. Ninety-five percent of our customers recommend us to their friends.
3. Nine out of 10 people who chewed our gum said it was the best banana-coconut gum they had ever chewed.
4. Our potato chips have 50% less fat.
5. This model is the most popular car in the central southwestern part of the state!

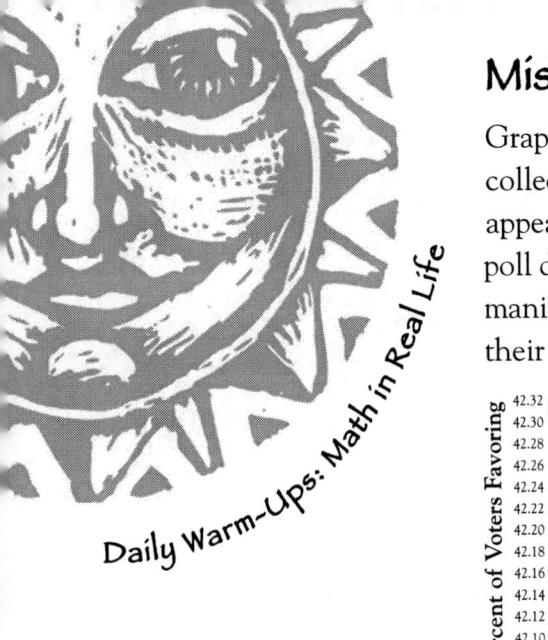

Misleading Graphs

Graphs organize large amounts of numbers or find patterns in data collections. Some graphs have been manipulated to make them appear more convincing than they are. Politicians manipulate poll data, scientists manipulate experiment results, and advertisers manipulate product information to try to convince you to believe their claims.

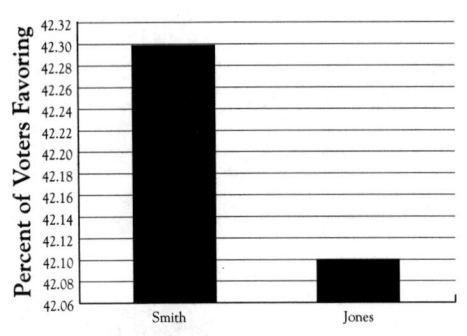

The chart above clearly shows that Smith has a staggering lead over Jones. Upon closer inspection, you may find that something is wrong.

1. What is misleading about this graph?
2. Find another misleading graph in a newspaper or on the Internet.

Pizza Deal

Which is the better deal—a large pizza for $15.00 or two medium pizzas for $7.50 each? The best way to find out is to calculate the number of square inches of pizza you are getting for the amount of money you are paying. If two medium pizzas have more square inches than the large, then buying two mediums is a better deal. If they have less, then the large pizza is a better deal.

Look at the choices below and determine which is the best deal.

a. a large 18-inch pizza with any four toppings for $19.95
b. two medium 10-inch pizzas with any four toppings for $18.95
c. a super large 24-inch pizza with any four toppings for $22.50
d. four small 6-inch pizzas with any four toppings for $17.95

Distance Between Points

There are many ways to measure distance. There is the straight distance between two points, or the distance traveled by the path of a car traveling between two locations. The distance traveled by a car is almost always longer than the straight-line distance between the two places.

You go on a road trip and travel 575 miles according to the car's odometer. When you use a ruler to measure the straight-line distance in your atlas, however, you note a distance of 512 miles.

Answer the following.

1. What is the difference between the distance measured with the odometer and the distance shown in the atlas?

2. List five things that could occur during a car trip that could account for the difference in the two distances.

3. What form of transportation allows you to travel in a straight line?

Rate

Rate is a measure of something that happens in a given amount of time. The number of miles traveled in an hour, the number of sodas bottled in a factory in a day, and the number of times the earth rotates in a year are all rates. Rate can be calculated by dividing the events or number of items by the time in which the events happen or the items are counted.

Answer the following questions.

1. In one factory, 50,000 toothpicks were made in 8 hours. What is the rate in toothpicks per day?

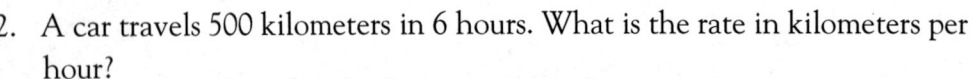

2. A car travels 500 kilometers in 6 hours. What is the rate in kilometers per hour?

3. A factory makes 150,000 gallons of gasoline in a day. What is the rate of production in gallons per year?

4. A farmer on a giant goat farm sells two goats every day. What is the rate of goats sold in goats per week?

Lightning Crashes

When lightning strikes close enough for you to see it, the light reaches you at a speed of 186,000 miles per second. Even if the lightning strike was 10 miles away, the light would reach you in 0.00005 seconds. The sound from the thunder travels at a much slower speed, about 1130 feet per second. This means that for every second between the time you see the lightning and hear the thunder, the sound has traveled 1130 feet.

Answer the following questions.

1. How many feet away was a lightning strike that produced thunder 4 seconds later at your location?

2. How many feet away was a lightning strike that produced thunder 8 seconds later at your location?

3. How many miles away was a lightning strike that produced thunder 20 seconds later at your location?

4. How many miles away was a lightning strike that produced thunder 5 seconds later at your location?

Megapixel Madness

There are a wide variety of digital cameras on the market. One of the first things a person might want to know when buying a digital camera is the number of megapixels the camera is capable of using to capture an image. A pixel is the smallest single point of light of a digital image or screen. The more pixels in a picture, the more information the camera has captured. This makes the image easier to magnify and to modify after it is taken. A megapixel is 1 million pixels. The number of megapixels can be found by multiplying the horizontal and vertical widths of an image (in pixels) and then dividing the total by 1 million.

Determine the number of megapixels in each digital image described below.

1. 3600 pixels × 2400 pixels

2. 2048 pixels × 1536 pixels

3. 640 pixels × 480 pixels

4. 4992 pixels × 3328 pixels

5. 2592 pixels × 1944 pixels

6. 1600 pixels × 1200 pixels

How Loud?

A *decibel* is a unit for measuring the general level of loudness. The calculation of the loudness is logarithmic. This means that for every increase by a factor of 10 decibels, the average person would perceive that the sound has doubled in loudness. Below is a chart of some common events and their decibel levels. Use the chart to answer the questions that follow.

Decibel level (dB)	Examples
30	whisper
40	mosquito buzzing
50	normal conversation
70	vacuum cleaner
80	alarm clock
120	threshold of pain
130	thunderclap

1. How many times louder is a mosquito buzzing than a whisper?

2. How many times louder is a thunderclap than normal conversation?

3. How many times quieter is a vacuum cleaner than the threshold of pain?

Calories per Serving

If you look quickly at the nutritional information on a 24-ounce bottle of soda, you might see that the soda contains 110 calories. If you read more carefully, however, you will see that a 24-ounce bottle actually contains three servings. This means the whole soda contains 330 calories. Soda is packed with calories, and the average teenager in the United States drinks the equivalent of somewhere between 700 and 875 cans of soda per year at about 165 calories per can.

Answer the following questions.

1. You can lose a pound of weight by burning about 3500 calories. How many cans of soda is that?

2. If you drank the equivalent of 875 cans of soda, how much weight could you gain?

3. A king-size candy bar has 2.5 servings, and a single serving has 280 calories. How many calories are in the whole candy bar?

4. How many king-size candy bars would it take for you to get a pound's worth of calories? (Remember, a pound is about 3500 calories.)

Ice Water Calories

You can lose weight by drinking ice water. Seems too good to be true? If you drink a 1-liter bottle of ice water, your body eventually heats the water up to your normal body temperature. This takes energy. In food calories, the amount of energy needed to raise a liter of water (33.8 ounces) from 0°C (possible ice water temperature) to 37°C (average human body temperature) is 1 calorie for each liter of water times the number of degrees the temperature goes up.

Solve each problem below.

1. How many calories does your body burn by raising the temperature of 1 liter of ice water from 0°C to 37°C?

2. Eight glasses of ice water is approximately 2 liters. If you drank 8 glasses of ice water in a day, how many calories would this burn?

3. You have to burn about 3500 calories to lose a pound. If you drink 2 liters of ice water every day, how many days would it take you to lose a pound?

Burning Calories

The chart below shows the number of calories burned when a person of a certain weight walks 1 mile at different speeds. So to burn 65 calories, a 100-pound person would have to walk 1 mile at 2 miles per hour. Use the chart to answer the questions that follow.

Weight in pounds	100	120	140	160	180	200	220
Speed walked	\multicolumn{7}{c}{Calories Burned}
2 mph (30-minute mile)	65	80	93	105	120	133	145
3 mph (20-minute mile)	60	72	83	95	108	120	132
4 mph (15-minute mile)	59	70	81	94	105	118	129
5 mph (12-minute mile)	77	92	108	123	138	154	169

1. How far would a 160-pound person have to walk at 3 miles per hour to burn 380 calories?

2. How many calories would a 220-pound person burn if he or she walked 6 miles at 4 miles per hour?

3. How far would a 180-pound person have to walk at 5 miles per hour to burn 1000 calories?

4. How many calories would a 100-pound person burn if he or she walked 8 miles at 2 miles hour?

Caffeine

Caffeine is a chemical commonly found in coffee, soda, and energy drinks, in pills for staying alert, and in medicine. A cup of regular coffee generally has about 90 milligrams of caffeine. A cup of tea has about 35 milligrams, a cup of hot chocolate has about 5 milligrams, a 1.5-ounce chocolate bar has about 9 milligrams, and a caffeinated soda has about 35 milligrams of caffeine.

Solve each problem below.

1. How many chocolate bars would you have to eat to get the same amount of caffeine that you could get from a cup of coffee?

2. How many cups of hot chocolate would you have to drink to get the same amount of caffeine as you would find in a caffeinated soda?

3. How many sodas would you have to drink to get as much caffeine as you would get in a cup of coffee?

4. You start each day with 4 servings of coffee. How many cups of tea would you have to drink to get the same amount of caffeine?

5. How many milligrams of caffeine would you get every day if you drank 8 cups of coffee?

Half-Life

The half-life of a radioactive material is the amount of time it takes for one half the material to decay and disappear. Some biological and chemical substances have a half-life as well. For example, if you have 200 grams of a material that has a half-life of 4 days, after 4 days you would have 100 grams left. After 8 days you would have 50 grams left. After 12 days you would have 25 grams left. After 16 days you would have 12.5 grams left, and so forth.

Solve each problem below.

1. A sample of plutonium has a half-life of 88 years. If you have a 350-gram sample, how much will be left after 352 years?

2. You work at a medical research facility and discover that a chemical with a half-life of 5 days has spilled. If 1000 milliliters spilled, how much remains after 30 days?

3. You find 60 grams of a medicine that has a half-life of 10 days. How much medicine was in the original sample if it has been out of its container for 100 days?

Just Average

Your final grade in a class is an average, or weighted average, of all your individual grades in a class. To find the average in any class, you need to divide the total number of points you earned by the total number of points that you could have earned. For example, 1129 points earned out of a possible 1345 points in a quarter means a grade of 84 (found by dividing 1129 by 1345).

Solve each problem below.

1. A student's grades were 77, 21, 90, 98, 96, 95, 91, 84, 88, 100, 100, 89, 67, 100, 98, 100, 64, 80, 99, 99, 99, 95, and 97. What is her average?

2. A student's grades were 64, 23, 54, 44, 46, 81, 72, 73, 82, 45, 89, 73, 83, 73, and 52. His teacher dropped the two lowest grades before taking the average. What is his average?

3. Your grades for a month are as follows—test grades: 100, 95, 97; quiz grades: 80, 85, 70; homework grades: 100, 95, 92, 89, 65, 98. If your teacher for this class counts each test grade twice, what is your average?

Sports Standings

If you are familiar with baseball, you know that there are a lot of statistics connected to the game. Knowing all the stats can be fun, but by the end of the season it really comes down to how well your team is doing in the standings. To find how many games behind first place a certain team is, you subtract that team's wins from the first-place team's wins. Then you add the lower-place team's losses and subtract the first-place team's losses. When you get your answer, divide by 2 to find out the number of games behind first place the lower-place team is.

Below are some standings from the American League East in Major League Baseball. Complete the chart by determining the number of games behind first place each team is. The first place team is, of course, zero games behind first place.

East	W	L	PCT	GB
Boston	61	40	.604	—
New York	60	40	.600	
Toronto	57	46	.553	
Baltimore	47	57	.452	
Tampa Bay	42	61	.408	

Daily Warm-Ups: Math in Real Life

68

© 2007 Walch Publishing

Saving on Heating Bills

It can be very expensive to heat a home when it's cold outside. Many people try various strategies to lower the amount of fuel they use to heat their homes. In most cases, these efforts save both money and the environment.

For each case below, determine the original cost for heating per year.

1. Installing an automatic thermostat (to reduce the temperature at night) that costs $120.00 cuts the heating bill by 8% and will pay for itself in 2 years.

2. Covering windows with plastic that costs $47.00 cuts the heating bill by 5% and will pay for itself in 1 year.

3. Insulating the attic costs $650.00. It cuts the heating bill by 20% and will pay for itself in 3 years.

Powers and Exponents

A number such as 4^3 has a base number, which is 4, and an exponent, which is 3. A number that is shown as a combination of a base and an exponent is called a power. Powers are a shorter way of writing some longer expressions.

Example: $4 \times 4 = 4^2$ $d \cdot d \cdot d \cdot f \cdot f \cdot f \cdot f = d^3 f^4$

Solve each problem below.

1. You start a penny collection. You have 2 pennies the first day, and the number doubles every day for 5 days. What is the expression with an exponent that represents the number of pennies you have?

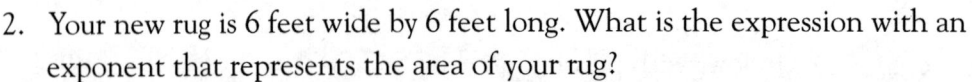

2. Your new rug is 6 feet wide by 6 feet long. What is the expression with an exponent that represents the area of your rug?

3. A video game console comes in a 65 cm by 65 cm by 65 cm box. What is the expression with an exponent that represents the volume of the box?

4. The number of blogs doubles every 6 months for $3\frac{1}{2}$ years. What is the expression with an exponent that represents the number of blogs at the end of that time period?

Square Roots

A factor that you multiply by itself to get a number is called the *square root* of that number. A *perfect square* is a number whose square root is a whole number. For example, 100 is a perfect square. The square root of 100 is 10 (and −10). The mathematical symbol for a square root is called a radical and is used in the following example: $\sqrt{100} = 10$

Solve each problem below.

1. You are making a square awning to shade people during a backyard barbeque. You buy 100 square feet of canvas. How long will each side of your awning be?

2. You decide your awning will not be large enough, so you go back and buy another 100 square feet of canvas for a total of 200 square feet. How long will each side of your awning be?

3. Your lawn is a perfect square and covers 1000 square feet. How long is your lawn on each side?

Least Common Multiple

The least common multiple of two or more numbers can be found by listing all the multiples of both numbers and finding the one of lowest value, excluding zero. Find the least common multiple in each situation below.

1. You are making sandwiches for a fund-raiser. Each loaf of bread has 20 slices, and each package of sandwich meat has 8 slices. If you use 2 slices of bread and 1 slice of meat in each sandwich, what's the least number of packages you have to buy so there are no leftovers?

2. You are buying hot dogs for a barbeque. Hot dogs come in packages of 10, and the buns come in packages of 8. What's the least number of packages you have to buy in order to have the same number of hot dogs and buns?

3. At the hardware store, you are looking for nuts and bolts for a building project. Nuts come in packages of 8, and bolts come in packages of 12. What's the least number of packages you have to buy in order to have the same number of nuts and bolts?

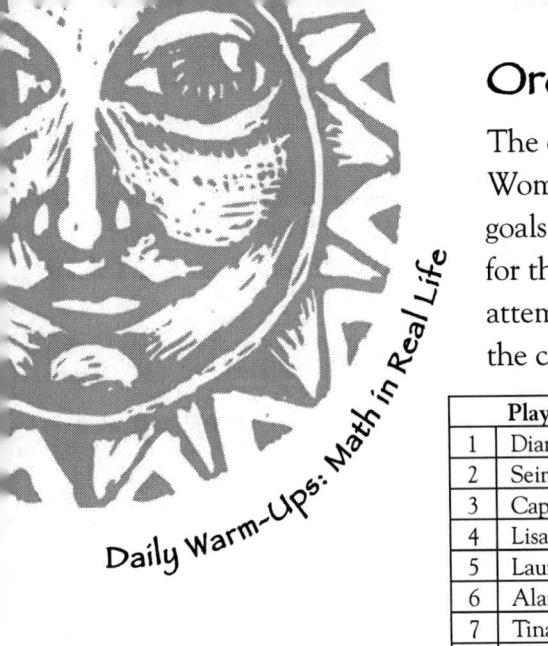

Ordering Decimals

The chart below shows statistics for some of the best players in the Women's National Basketball Association. *FGM* stands for field goals made, and *FGA* stands for field goals attempted. *3PM* stands for three-point shots made, and *3PA* stands for three-point shots attempted. The decimals signify that the numbers are averages. Use the chart to complete the questions that follow.

	Player name	FGM–FGA	FG%	3PM–3PA	3P%
1	Diana Taurasi	8.3–18.8	.441	3.3–8.6	.384
2	Seimone Augustus	8.4–18.0	.465	0.8–2.5	.338
3	Cappie Pondexter	7.2–16.2	.441	1.7–4.4	.377
4	Lisa Leslie	7.5–14.3	.528	0.3–0.5	.533
5	Lauren Jackson	6.4–12.0	.537	1.0–2.6	.406
6	Alana Beard	7.0–14.2	.491	0.9–2.8	.324
7	Tina Thompson	6.2–13.7	.450	1.8–4.4	.422
8	Katie Douglas	5.6–12.6	.445	2.5–5.6	.441
9	Tamika Catchings	4.8–12.6	.379	1.1–3.7	.298
10	Tamika Whitmore	5.7–12.7	.447	0.4–0.9	.435

1. Put these players in order from highest percent of FGM to lowest percent of FGM.

2. Put these players in order from highest percent of 3PM to lowest percent of 3PM.

Tire Pressure

If you ever have to inflate a tire, you should know that it can be dangerous to put too much air into the tire. Just like a balloon, a tire can explode if the pressure inside gets too great. Tire pressures are generally measured in pounds per square inch (psi) or kilopascals. When you read a tire gauge, you are seeing the combination of the pressure inside the tire pushing out and the pressure outside the tire pushing in. The formula to find absolute pressure is absolute pressure = gauge pressure + atmospheric pressure. *Absolute pressure* is all the pressure in the tire, *atmospheric pressure* is supplied by the atmosphere, and *gauge pressure* is what the tire gauge reads.

Solve each problem below.

1. The absolute pressure in your tire is 54.7 psi, and your tire gauge reads 40 psi. What is the atmospheric pressure?

2. You need to fill your tires to 30 psi. The atmospheric pressure is 14.7 psi. What is the absolute pressure in the tire?

3. The absolute pressure in your tire is 50 psi. The atmospheric pressure is 14.7 psi. What does your tire gauge read?

Multiplying with Decimals

Many of the things we use daily have decimal values associated with them. Stopwatches have a decimal measure for tenths and hundredths of a second. A purchase that doesn't come out to an exact dollar has a decimal in the form of cents. Purchases made by weight at the store are seldom exactly a pound.

Solve each problem below.

1. You buy 0.74 pounds of macaroni salad on sale for $.79 per pound. What is the cost?

2. You need a rug to cover a space in your living room that is 77.5 inches by 80.25 inches. How many square inches should the rug cover?

3. Your gas tank holds 14.2 gallons, and gasoline costs $3.15 a gallon. What will it cost to fill up your gas tank?

4. Your water bottle holds 16.9 ounces of water. You drink 2.5 ounces. How many ounces of water are left?

Dividing with Decimals

When you have a decimal number, you may need to determine how many parts it can be split into. When you are working with percents or fractions, you can change the percent or fraction into a decimal before you divide.

Solve each problem below.

1. You have $\frac{3}{4}$ of a pizza that is split into 7 slices. How much of the original pizza is represented by each slice?

2. You buy .90 pounds of carrots to serve to 4 people. How much will each person get if the carrots are split into 4 equal portions?

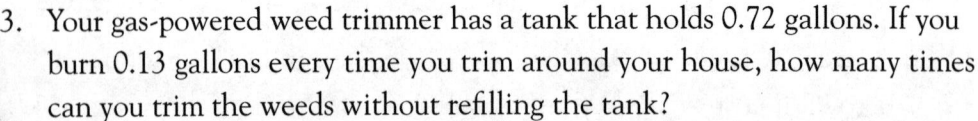

3. Your gas-powered weed trimmer has a tank that holds 0.72 gallons. If you burn 0.13 gallons every time you trim around your house, how many times can you trim the weeds without refilling the tank?

4. You have 0.54 meters of string to use for a homework project. You need to cut it into pieces that are 0.08 meters each. How many whole pieces can you get?

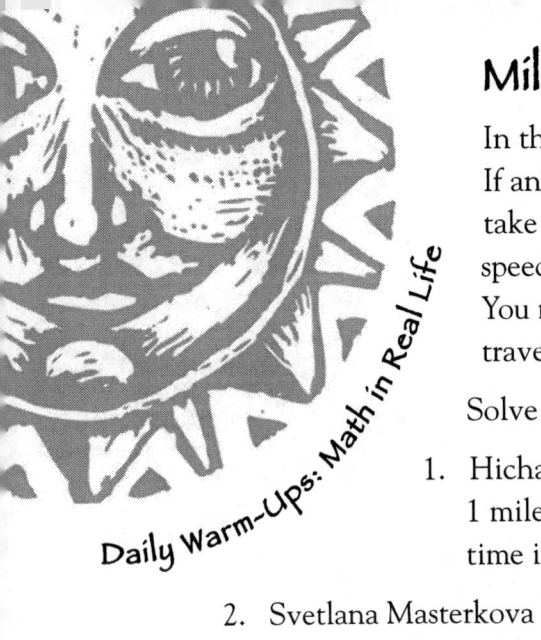

Miles per Hour

In the United States, speed is usually measured in miles per hour. If an object is traveling at 1 mile per hour, it means that it would take the object 1 hour to travel 1 mile. Cars routinely travel at speeds as high as 75 miles per hour legally on some U.S. highways. You might be surprised to discover just how slowly human beings travel when you convert their speeds to miles per hour.

Solve each problem below.

1. Hicham El Guerrouj of Morocco set a world record by running 1 mile in 3 minutes, 43.13 seconds on July 7, 1999. What was his time in miles per hour?

2. Svetlana Masterkova of Russia set a record when she ran 1 mile in 4 minutes, 12.56 seconds. What was her time in miles per hour?

3. Paul Tergat set a record by running 26.22 miles in 2 hours, 4 minutes, and 55 seconds on September 28, 2003. What was his time in miles per hour?

4. Paula Radcliffe set a world record on April 13, 2003 when she ran 26.22 miles in 2 hours, 15 minutes, and 25 seconds. What was her time in miles per hour?

Miles per Gallon

When you are shopping for a vehicle, you might want to consider what the vehicle gets for gas mileage. From compact cars to SUVs, there are a broad range of sizes to choose from. You could save a lot of money in the long run by purchasing a car that gets great mileage.

Determine the miles per gallon achieved by each of the vehicles described below.

1. A Ford Mustang can travel 448 miles on one 16.0-gallon tank of gas.

2. A Hummer H2 uses 32.0 gallons to go 307.2 miles.

3. The Honda Insight can go 198 miles on 3 gallons of gas.

4. A Toyota RAV4 can travel 477 miles on one 15.9-gallon tank of gas.

5. The BMW M5 can go 198 miles on 11 gallons of gas.

6. The Cadillac Escalade has a 26.0-gallon tank and can travel 468 miles before filling up again.

The Road Trip

Going on a family vacation, traveling with a group of friends, or just traveling by yourself can require a lot of money. Below is a short description of Jamal's trip driving from Bangor, Maine to Orlando, Florida. Read the description and then answer the question that follows.

Jamal drove a total of 3070 miles in a car that gets 32.4 miles per gallon. He was able to buy gasoline for $2.959 per gallon by pre-buying with a gas card. He stayed in hotels for a total of 11 nights with an average cost of $79.85 per night. His average breakfast cost $4.50, his average lunch cost $12.62, and his average dinner cost $23.98. He bought souvenirs for himself and his family totaling $225.60. He went to three theme parks for a total of $459.87 and rented a boat for a day for $239.95. He had purchased a digital camera for $202.16 for the trip and took 150 pictures that he had printed when he came home for $0.29 each. What was the total cost of Jamal's trip?

Speed Reading

If you are shopping for a new computer, you will see that a CD or DVD drive often has a number with an X after it, such as 16X or 48X, to describe its speed. For example, a 16X CD spins 16 times faster than a 1X CD. But just how fast is that? If you do not know what 1X is, you cannot determine how fast 16X is. For a CD, 1X means that it reads information at 176 kilobytes per second. For a DVD, 1X means that it reads 1.32 megabytes per second.

Determine the maximum read speed for each drive described below. List your answers in megabytes per second. (1000 kilobytes = 1 megabyte)

1. a 16X CD
2. a 32X CD
3. a 52X CD
4. a 4X DVD
5. an 8X DVD
6. a 12X DVD

DVD versus CD Storage

A standard CD that you can burn music or data on usually has 700 megabytes (MB) of storage. DVDs have 4.7 gigabytes (GB), 8.5 gigabytes, or 9.4 gigabytes of storage. Remember that a gigabyte is made of 1000 megabytes. By way of comparison, a standard 700-MB CD can hold around 125 MP3 music files (depending on recording rate and length), while a 4.7-GB DVD can hold around 850 MP3 music files.

Solve each problem below.

1. How many CDs of music could you fit on a 4.7-GB DVD?

2. How many CDs of music could you fit on an 8.5-GB DVD?

3. How many CDs of music could you fit on a 9.4-GB DVD?

4. If you have to back up the 200-GB hard drive on your computer, how many 9.4-GB DVDs will it take?

5. If you have to back up the 200-GB hard drive on your computer, how many 700-MB CDs will it take?

Writing Decimals and Fractions

Fractions and decimals are both used to show parts of a whole. The fraction $\frac{1}{2}$, for example, means one part out of a whole split into two parts. A decimal such as 0.75 is another way of showing part of a whole. In this case, the whole number 1 has been split, so to speak, into 100 parts. The decimal 0.75 represents 75 parts out of 100, or $\frac{75}{100}$. We often use fractions when discussing measurement, and we use decimals almost every day when handling money or making purchases.

Convert each fraction below into a decimal, and convert each decimal into a fraction.

1. You buy a pack of gum for $.75.
2. A friend eats $\frac{5}{8}$ of your pizza.
3. You have done 0.5 of your homework.
4. You buy some shoes that are $\frac{2}{3}$ of the original price.
5. The temperature is 0.25° warmer than it was 5 minutes ago.
6. You have put together $\frac{21}{134}$ of your model airplane.

Building with Fractions

When making home improvements, you may choose to do the work yourself to save money. Standard rulers and tape measures are marked in feet and inches. The inches are split into halves, quarters, eighths, sixteenths, and sometimes thirty-seconds. To make something fit exactly, measure as carefully as you can. When you are multiplying or adding the fractions, the more accurate your measurements, the more accurate your final product will be.

Solve the following measurement problems.

1. You are putting hardwood flooring into a room in your home that is 12 feet, $6\frac{1}{4}$ inches long and 10 feet, $4\frac{1}{2}$ inches wide. What is the exact space that will be covered by the hardwood flooring?

2. A new kitchen countertop has to fit in a space that is 9 feet, $3\frac{1}{4}$ inches long. There is a piece of trim on either end that is $\frac{7}{8}$ of an inch wide. How long should the countertop be to fit between the two trim pieces?

3. The shelves in your new bookcase are $36\frac{1}{8}$ inches long. The case on either side is $\frac{3}{4}$ of an inch. What is the total width of your bookcase with the shelf in it?

Multiplying by Fractions

Not every item you buy has to be bought in round numbers of units. For example, if you go to the deli to buy some sliced turkey breast, you might buy less than a pound if you only want enough for a few sandwiches. Likewise, if you are making something out of fabric, you might only need $3\frac{1}{2}$ feet of fabric instead of a full 4 feet.

Solve each problem below.

1. You want to buy sliced ham at the deli, and it is $3.99 a pound. What will $\frac{1}{4}$ of a pound of ham cost?

2. You go to the fabric store to buy some denim. Fabric is $1.59 per yard and you need 6 feet, 4 inches of fabric. How much will the fabric cost?

3. You have a recipe for cookies and it calls for $\frac{3}{4}$ of a cup of flour. How many cups of flour will you need if you are going to make 8 batches of cookies?

4. You are buying a piece of carpeting that has to cover an area that is 6 feet, $4\frac{1}{2}$ inches by 4 feet, $3\frac{1}{8}$ inches. How much carpeting will you have to buy?

Dividing by Fractions

Sometimes you need to know how many parts are in a whole. To find out, you may need to divide by a fraction. For example, say you have a 500-foot coil of rope. If you need to know how many $\frac{3}{4}$-foot pieces of rope could be cut from it, you would divide the whole number 500 by the fraction $\frac{3}{4}$.

Solve each problem below.

1. You have an 8-foot board. How many smaller pieces can you cut out of the board that are $1\frac{1}{8}$ inches long?

2. You are making mini apple pies, and each pie requires $\frac{2}{3}$ of an apple. If you have 62 apples, how many mini apple pies can you make?

3. You are making dinner and have a 5-pound package of ground beef. How many quarter-pound hamburgers can you make?

4. You are putting chain link fence around a garden plot. If the total distance around the plot is 150 feet and the link fence comes in $3\frac{3}{4}$-foot lengths, approximately how many sections of fencing will you need?

Average Speed

The average speed of an object can be found by taking the distance the object traveled and dividing it by the time it took to travel that distance. For example, a trip covering 100 miles might include stops at rest areas, gas stations, toll booths, and restaurants. The clock keeps ticking even when the car isn't moving. Every second you sit still, your average speed is decreasing.

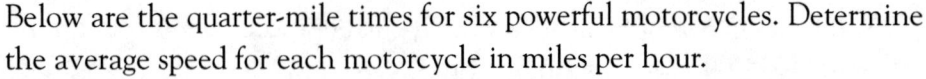

Below are the quarter-mile times for six powerful motorcycles. Determine the average speed for each motorcycle in miles per hour.

1. The BMW XV12R traveled $\frac{1}{4}$ mile in 11.43 seconds.
2. The Ducati 999R traveled $\frac{1}{4}$ mile in 10.10 seconds.
3. The Honda CBR1000RR traveled $\frac{1}{4}$ mile in 9.89 seconds.
4. The Kawasaki ZX-10R traveled $\frac{1}{4}$ mile in 9.78 seconds.
5. The Suzuki GSX-R600 traveled $\frac{1}{4}$ mile in 11.20 seconds.
6. The Yamaha YZF-R6 traveled $\frac{1}{4}$ mile in 10.54 seconds.

Metric Fractions

One of the advantages of using the metric system (or more accurately the International System of Units) is that you can convert from unit to unit by multiplying or dividing by 10. This means that in general you can use simple fractions when performing mathematical operations such as adding, subtracting, multiplying, and dividing.

For each statement below, write a fraction that represents the information.

1. There are 10 millimeters in 1 centimeter.
2. There are 100 millimeters in 10 centimeters.
3. There are 1000 millimeters in 1 meter.
4. There are 1,000,000 micrometers in 1 meter.
5. There are 100 micrometers in 1 millimeter.
6. There are 1000 meters in 1 kilometer.

Daily Warm-Ups: Math in Real Life

Ratios

A ratio is a way of comparing two numbers by division. For instance, if you had 16 sneakers and 8 boots, the ratio of sneakers to boots would be written as 16:8. The ratio can also be written as $\frac{16}{8}$. In this case, the fraction $\frac{16}{8}$ can be reduced to 2. The ratio of sneakers to the total number of items would be 16:24, which can be read as "sixteen to twenty-four" or "sixteen out of twenty-four."

Solve each problem below.

1. You have 12 T-shirts and 6 pairs of jeans. What is the ratio of jeans to T-shirts?

2. You have 4 people in your family and 2 cars. What is the ratio of family members to cars?

3. You and 5 friends sit down to enjoy 2 pizzas that have each been cut into 8 slices. What is the ratio of slices to people?

4. You spend 3 hours studying. What is the ratio of hours studying to hours in a day?

5. You cook a pie that has $1\frac{1}{2}$ cups of chocolate chips for every $2\frac{1}{2}$ cups of flour. What is the ratio of chocolate chips to flour?

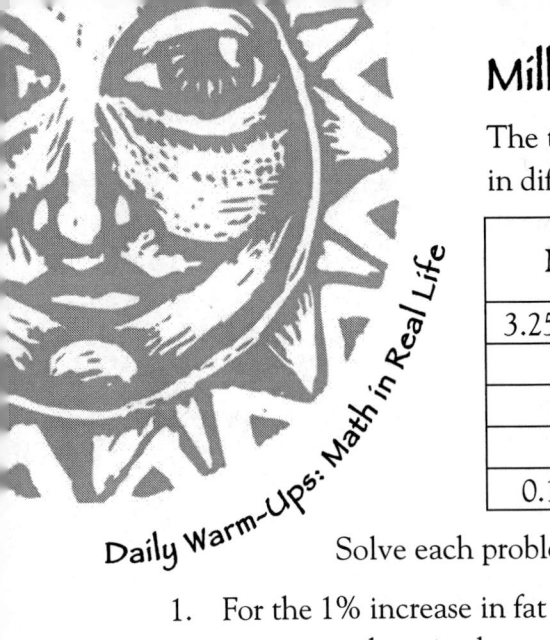

Milk Fat

The table below shows the total calories and the calories from fat in different kinds of milk.

Milk fat	Calories per 8 ounces	Calories from fat
3.25% (whole)	150	70
2%	130	45
1.5%	120	35
1%	110	20
0.1% (skim)	90	0

Solve each problem below.

1. For the 1% increase in fat between 1% and 2% milk, how much of an increase is there in the number of calories? In the number of calories from fat?

2. How do the calories from fat in 1.5% milk compare with the calories from fat in whole milk?

3. How many calories are there in 1 gallon (128 ounces) of whole milk?

4. How many calories are there in 1 gallon (128 ounces) of skim milk?

Free Throws

If you play basketball or have ever watched a game, you are probably familiar with the idea of a free throw, which is also often called a foul shot. Some basketball players have made a career out of being a great shot from the foul line, while other players struggle.

Player	F.T. Made	F.T. Attempted	%
Mark Price	2135	2362	
Peja Stojakovic	1864	2086	
Steve Nash	1726	1926	
Rick Barry	3818	4243	
Calvin Murphy	3445	3864	
Shaquille O'Neal	5147	9744	
Michael Jordan	7327	8772	

Answer the following using the chart above.

1. Determine the percentage of free throws made for each player.

2. Who is the best free-throw shooter on the list?

3. Who is the worst free-throw shooter on the list?

4. How did Michael Jordan compare to Mark Price at free throws?

Winning Percentage

Below are some win/loss records from the American Football Conference in the National Football League. To find the winning percentage for each team, take the number of wins and divide it by the total number of games played.

Team	W	L	W%
New England	10	6	
Miami	9	7	
Buffalo	5	11	
N.Y. Jets	4	12	
Cincinnati	11	5	
Pittsburgh	11	5	
Baltimore	6	10	
Cleveland	6	10	

Team	W	L	W%
Indianapolis	14	2	
Jacksonville	12	4	
Tennessee	4	12	
Houston	2	14	
Denver	13	3	
Kansas City	10	6	
San Diego	9	7	
Oakland	4	12	

Answer the following using the charts above.

1. Calculate the winning percentages for all the teams.
2. Which team had the highest winning percentage?
3. Which team had the lowest winning percentage?

Recreation Time

There are only 24 hours in a day, and many of those hours are spent sleeping. When school is in session, you probably spend 6 to 8 hours doing school-related activities (even more if you take part in extracurricular activities, such as sports and clubs). The question then becomes, how much free time is left for fun?

Answer each question below. Imagine that you spend 8 hours a day sleeping and 8 hours at school or getting to and from school.

1. What percent of your free time is taken up by a dinner that lasts 1 hour and 15 minutes?

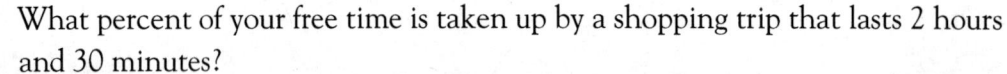

2. What percent of your free time is taken up by a shopping trip that lasts 2 hours and 30 minutes?

3. After 2 hours of homework, 3 hours of television, and 2 hours of chatting with friends online, what percent of your free time remains?

4. Think of some of the activities that you do in your free time. What percent of your time does each of these activities take up?

Extraordinary Percents 1

Some percents are greater than 100%. Percents greater than 100% indicate that a comparison is being made. For example, say the recommended daily allowance (RDA) of vitamin C is 90 milligrams. If you consume 180 milligrams in one day, then you've taken 200% of the recommended dose.

Solve each problem below.

1. The RDA of protein for an average-size man is about 60 grams a day. What percent of the normal RDA would an average-size man get if he ate 100 grams in a day?

2. Each morning you normally run 4 miles. This morning you ran 5 miles. What percent of your daily distance did you complete compared to normal?

3. The average 5-year-old child is 44 inches tall. Your 5-year-old cousin is 50 inches tall. What percent of average height is he?

4. The average income for a fast-food cook is $6.50 per hour. You get hired and are earning $8.00 per hour. What percent of the normal income are you earning?

Extraordinary Percents II

Some percents are less than 1%. Such small percentages are a way of expressing a very small amount of something compared to the total amount. For example, if your total take-home salary is $35,000 and you purchase a $1.00 soda, the percentage of your salary the soda represents is $1.00 ÷ $35,000 × 100 = 0.003%.

Imagine that you take a multivitamin tablet every morning that has a total of 1255 milligrams. Determine the percent of your vitamin represented by each of the nutrients below.

1. vitamin B-1: 50 mg
2. zinc: 25 mg
3. copper: 2 mg
4. inositol: 10 mg
5. silica: 4 mg
6. molybdenum: 0.75 mg

Percent Change

When a number increases, the amount that the number increases compared to what it was originally is a percent increase. If a number decreases, the amount of change is a percent decrease. The formula for percent change is written below.

$$\text{percent} = \frac{\text{amount of change}}{\text{original amount}}$$

Solve each problem below using the percent change formula.

1. You have 30 pretzels and give 4 away. What is the percent decrease?

2. At the track meet, you jump 17 feet in the long jump on your first attempt. You jump 17 feet, 9 inches on your second attempt. What is the percent increase in your distance?

3. You were 58 inches tall last year. Now you are 65 inches tall. What is the percent increase in your height?

4. A tree near your home was planted when it was 6 feet tall. Now the tree is 52 feet tall. What is the percent increase in the tree's height?

Steepness

Steepness is a comparison of how rapidly a vertical height increases over horizontal distance. For example, when driving along a stretch of highway, the road level might increase by 1 mile over the course of 20 miles. The rise of 1 mile divided by the run of 20 miles gives a number of .05 or a 5% measure of steepness, sometimes called grade.

Answer the following questions.

1. You are building a ramp for wheelchair access into a new building and the vertical height increases 1 inch for every 12 inches of horizontal distance. What is the steepness of the ramp as a percentage?

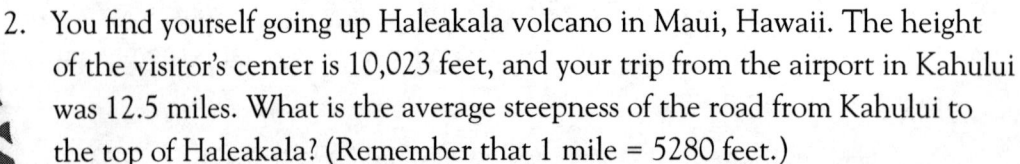

2. You find yourself going up Haleakala volcano in Maui, Hawaii. The height of the visitor's center is 10,023 feet, and your trip from the airport in Kahului was 12.5 miles. What is the average steepness of the road from Kahului to the top of Haleakala? (Remember that 1 mile = 5280 feet.)

3. Walking up the stairs in your home, you find that you go up 10 feet and travel a horizontal distance of 14 feet. What is the steepness of your stairs as a percentage?

Extrapolation

Extrapolation is the process of using existing data and estimating a value or values that go beyond the points in a set. For example, in the number sequence 2, 4, 6, 8, 10, 12, 14, ___, ___, you can probably tell that the missing numbers are 16 and 18. The prefix *extra-* means "beyond" or "outside." You use extrapolation when you are looking for missing numbers that are beyond numbers you already know.

Solve each problem below.

1. Your salary for the last five years has been $30,000, $33,000, $36,000, $39,000, and $42,000. What is your salary likely to be for the next three years?

2. The top speed of a new motorcycle is 175.6 miles per hour (mph). Over the last few years, the new models have had a top speed of 174.0 mph, 174.4 mph, 174.8 mph, and 175.2 mph. What is the likely top speed of the next new model?

3. The stock you own was worth $20 when you first bought it. The value over the next three years was $22, $24.20, and $26.62. What is the likely value of the stock at the end of this year?

Interpolation

Interpolation is the process of using existing data to estimate a value or values between points in the set. For example, in the number sequence 1, 2, 3, ___, 5, 6, 7, you can probably tell that the missing number is 4. The prefix *inter-* means "among" or "between." You often use interpolation when you are reading a line graph.

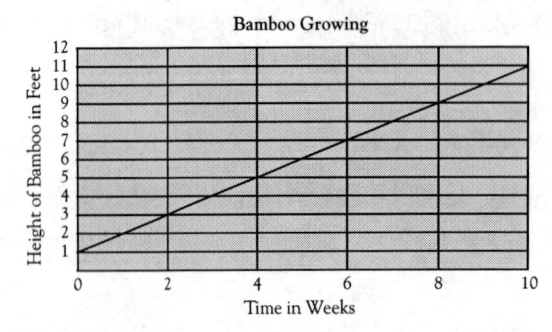

Bamboo Growing

The graph above represents the growth pattern of bamboo. Use the graph to answer the following questions.

1. How tall was the bamboo after 1 week?

2. How tall was the bamboo after 9 weeks?

3. When was the bamboo 4 feet tall?

4. When was the bamboo 8 feet tall?

English Length

Before the metric system was developed, the English system of units was commonly used. The English system is still used in the United States, Canada, and England, although these countries use the metric system also. In the English system, the common units of length are the inch, the foot, the yard, and the mile. Below are some of the relationships between the units of distance.

1 foot = 12 inches   1 yard = 3 feet

1 mile = 5280 feet   1 mile = 1760 yards

Solve each problem below.

1. You have a friend who is 6 feet, 5 inches tall. How tall is he in inches?
2. The top of Mount Everest is 29,028 feet above sea level. How many miles is that?
3. You run 35 yards and score a touchdown. How many feet is that?
4. You drive your car 14.4 miles. How many feet is that?
5. You drive your car 14.4 miles. How many inches is that?

English Volume

In the English system, the common units for measuring fluid volume are the ounce, the cup, the pint, the quart, and the gallon. Below are some of the relationships between the units of volume.

1 cup = 8 ounces   1 pint = 2 cups
1 gallon = 4 quarts 1 quart = 2 pints

Solve each problem below.

1. You buy a gallon of milk. How many cups can you pour from it?

2. You buy a 36-ounce can of tomatoes. How many pints is that?

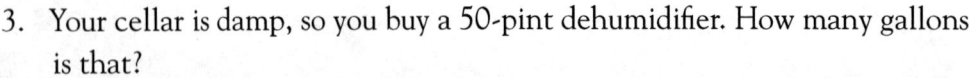

3. Your cellar is damp, so you buy a 50-pint dehumidifier. How many gallons is that?

4. To make a pint of iced tea, you need 3 tablespoons of iced tea mix. How many tablespoons would you need to make 3 quarts of iced tea?

5. Fertilizer for your lawn has to be mixed with water. Every 5 ounces of mix needs 1 gallon of water. How many cups of mix do you need if you are using 8 gallons of water?

English Weight

In the English system of units, the common units of weight are the ounce, the pound, and the ton. Below are some of the relationships between the units of weight.

1 pound = 16 ounces   1 hundredweight = 100 pounds

1 ton = 2000 pounds

Solve each problem below.

1. At your job you move 350 hundredweight barrels in one day. How many tons did you move?

2. You buy 5 pounds of nails at the store. How many ounces is that?

3. As the driver of an 18-wheeler, the maximum weight allowed for your truck in one state is 80,000 pounds. How many tons is that?

4. You drive your truck across a border and find that the weight limit in that state is 99,000 pounds. How many tons is that?

5. Your car weighs 1.722 tons. How many ounces is that?

Metric Length

The basic unit of length in the metric system is the meter. Different prefixes are put in front of the word *meter* to indicate different sized metric measurements. For example, a millimeter is about the thickness of a dime. A centimeter is about the height of a stack of five nickels. A meter is a little more than 3 feet. A kilometer is about 3281 feet, or about 0.62 miles.

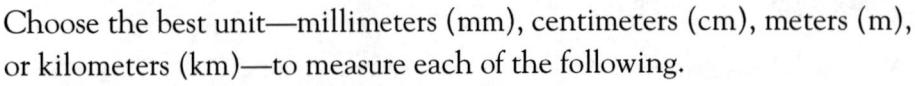

Choose the best unit—millimeters (mm), centimeters (cm), meters (m), or kilometers (km)—to measure each of the following.

1. your best friend's height
2. the thickness of a pepperoni slice
3. the distance from Chicago, Illinois to Melbourne, Australia
4. the height of the Statue of Liberty
5. the distance to Mars from Earth
6. the width of a highway lane

Metric Volume

The basic unit of volume in the metric system is the liter. You have probably seen a 1-liter bottle or a 2-liter bottle of soda. A milliliter is $\frac{1}{1000}$ of a liter, which means it takes 1000 milliliters to make a liter. A milliliter is about the size of ten to fifteen water droplets. You often see milliliters given as a unit of measure on medicines. A milliliter is also the same size as a cubic centimeter, or cc, which you might hear mentioned on medical shows or in a discussion about small engines.

Choose the best units—liters (l) or milliliters (ml)—to measure each of the following liquids.

1. the amount of water needed to fill an above-ground swimming pool
2. the amount of medicine that would fill half a teaspoon
3. the amount of water that would fill a large cooking pot
4. the amount of liquid that would fill a test tube
5. the amount of mouthwash that would fill a travel-sized container
6. the amount of coffee that would fill a thermos

Metric Mass

The metric system uses the kilogram as the base unit of mass. A kilogram is about 2.2 pounds, which is also the mass of 1 liter of water. Another common unit of mass is the gram, which is about the mass of a paper clip. A nickel has a mass of almost exactly 5 grams. A milligram is very small and would be about the mass of a piece of sand or a grain of salt. Below are some relationships between metric masses.

1000 milligrams = 1 gram   1000 grams = 1 kilogram

1 metric ton = 1000 kilograms

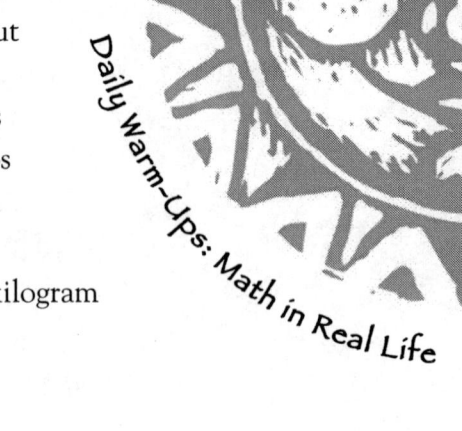

Solve each problem below.

1. You buy a liter of water. What is its mass in grams?

2. Your car has a mass of 2 metric tons. What is the car's weight in pounds?

3. You buy a 5-pound bag of sugar. What is its mass in grams?

4. The instructions on your aspirin bottle say take two 200-milligram aspirin. How many grams is the total dose?

Time

Both the metric system and the English system use the same basic units of time. Those units are the second, the minute, the hour, the day, and the year. There are also other common units of time such as months and weeks. Below are some of the relationships between the units of time.

1 minute = 60 seconds   1 hour = 60 minutes

1 day = 24 hours   1 year = 365 days

Solve each problem below.

1. You have 45 more days of school left. How many years is that?

2. Your job starts at 4:00 P.M. and ends at 8:30 P.M. How many minutes is that?

3. Light takes about 4.2 years to reach Earth from Proxima Centauri, which is the closest star to Earth other than the Sun. How many days does it take the light to travel that distance?

4. Calculate your age in hours to the nearest hour.

Reading Tables

In a physical science class such as chemistry or physics, information is often presented in a table format. Tables such as the one below are often found in the appendix of a chemistry text. You don't have to understand all the information presented in order to use the table.

Element	Symbol	Atomic number	Molar mass (g/mol)	Density (g/cm^3)	Melting point (°C)	Boiling point (°C)
argon	Ar	18	39.95	1.66	–189	–186
carbon	C	6	12.01	2.27	3700	—
gold	Au	79	196.97	19.28	1064	2807
iron	Fe	26	55.85	7.87	1540	2760
titanium	Ti	22	47.88	4.55	1660	3300

Answer the following about the chart above.

1. Which element has the highest density?

2. Which element has the highest melting point?

3. Which element has a symbol of Fe?

4. Which element has the lowest molar mass?

5. Which element has the highest boiling point?

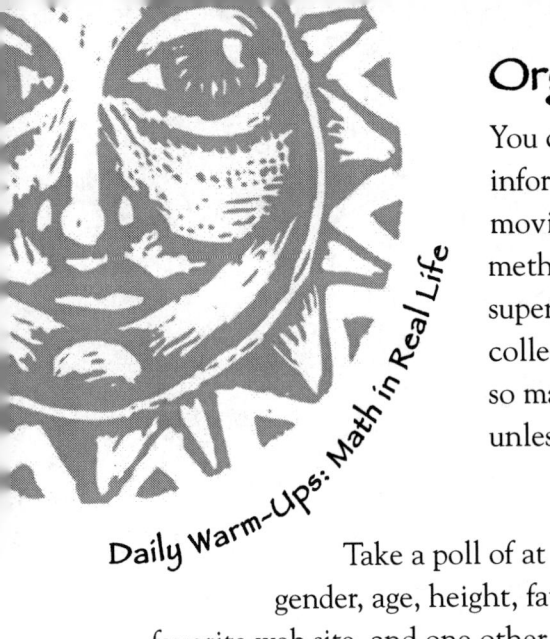

Organizing a Table

You can collect information in a number of ways. You might use an informal poll in which you ask a few friends, for example, "What movie do you want to go see?" You might use a more extensive method, such as if you were polling all the people who came into a supermarket to see what kind of produce they were buying. As you collect data, you will quickly find that you can only keep track of so many pieces of information, and that it is hard to find a pattern unless you organize the data to look at it.

Take a poll of at least 10 people and collect the following information: gender, age, height, favorite color, favorite movie, favorite snack food, favorite web site, and one other item of your choice. Organize your data so that it is in a single table, and then answer the questions below.

1. Is there a connection between gender and any other trend? Explain.

2. Is there a connection between age and any other trend? Explain.

3. Is there a connection between favorite color and favorite snack food? Explain.

Turning the Tables

A quick tour of any kitchen will reveal a number of items that contain nutritional information. For example, on a label of 1% milk, you can see that 1 cup (8 ounces) is considered a serving and should have about 110 calories, with 20 of those from fat. The rest of the label contains the following information: total fat 2.5 grams (4%), saturated fat 1.5 grams (8%), unsaturated fat 0 grams, cholesterol 15 milligrams (4%), sodium 125 milligrams (5%), total carbohydrate 13 grams (4%), dietary fiber 0 grams, sugars 12 grams, protein 8 grams (17%), vitamin A 10%, vitamin C 10%, calcium 30%, iron 0%, and vitamin D 25%.

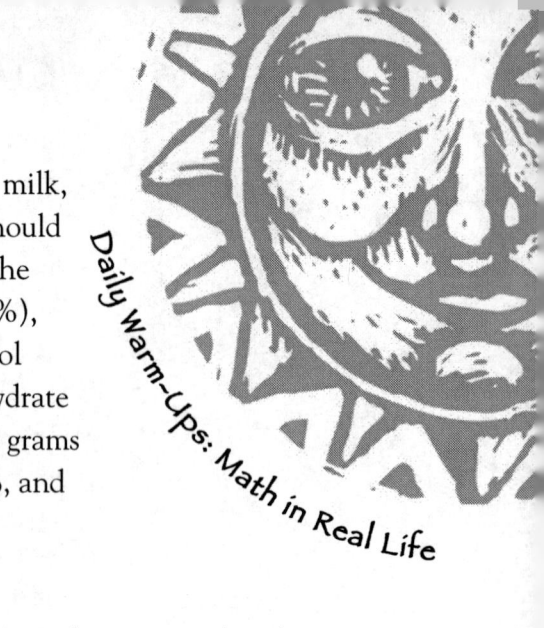

Think of two items that you might eat together, such as oatmeal and orange juice. Look at the nutritional information printed on the label or find the information online. Then create a table combining the nutritional information from both items.

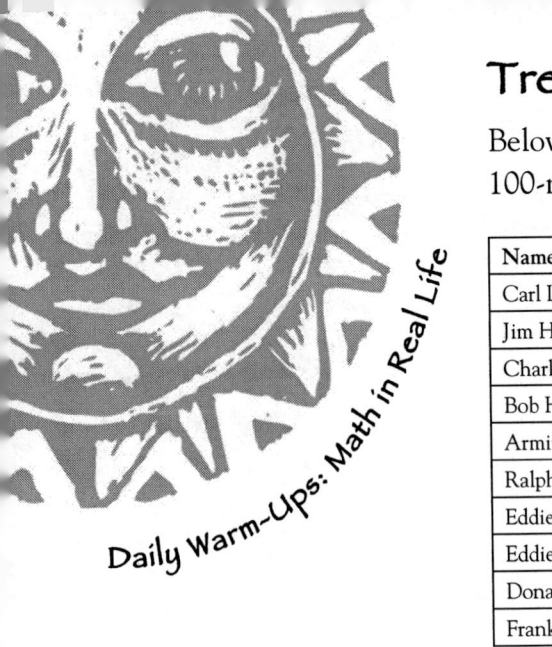

Trends and Records

Below is a table showing the Olympic record times for men in the 100-meter run between 1900 and 1988.

Name	Result	Date
Carl Lewis (USA)	9.92	Sept. 24, 1988
Jim Hines (USA)	9.95	Oct. 14, 1968
Charlie Greene (USA)	10.02	Oct. 13, 1968
Bob Hayes (USA)	10.0	Oct. 15, 1964
Armin Hary (FRG)	10.2	Aug. 31, 1960
Ralph Metcalfe (USA)	10.3	Aug. 1, 1932
Eddie Tolan (USA)	10.3	Aug. 1, 1932
Eddie Tolan (USA)	10.4	July 31, 1932
Donald Lippincott (USA)	10.6	July 6, 1912
Frank Jarvis (USA)	10.8	July 14, 1900

Answer the following questions about the table above.

1. Who held the record in 1940?
2. What was the longest that a record stood?
3. What was the largest difference in time between any two records?
4. Who broke his own record?

Reading Pie Charts

Pie charts appear on television, in newspapers and magazines, and in almost every form of visual media. A pie chart compares the parts of a whole to the whole. Below is a chart showing what pets people own in one town. Use the chart to answer the questions that follow.

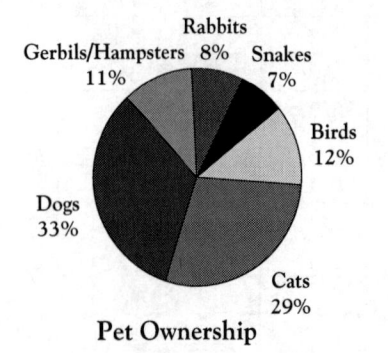

Pet Ownership

- Rabbits 8%
- Snakes 7%
- Birds 12%
- Cats 29%
- Dogs 33%
- Gerbils/Hampsters 11%

1. What percent of the pets in town are dogs?

2. What percent of the pets in town are cats?

3. Do the rabbits, snakes, and birds make up a larger percent than the cats?

4. What pets are least likely to be owned?

5. What is the difference in the percent of cats and birds owned?

Reading Bar Graphs

A bar graph is a good graph to use to count something or compare similar things to see how closely they relate to one another. The chart below is the result of a poll in which people were asked to choose the candidate they found the most trustworthy.

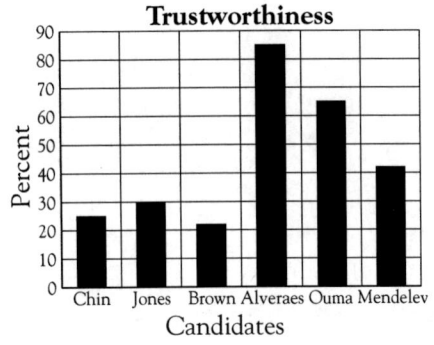

Answer the following questions about the chart above.

1. Which candidate seems the most trustworthy?

2. Which candidate seems the least trustworthy?

3. What percent of people thought Chin was trustworthy?

4. What percent of people thought Ouma was trustworthy?

5. What candidate was considered 85% trustworthy?

Making a Bar Graph

Bar graphs are a great tool for displaying information that you have counted. For example, if you counted the number of red, blue, and green cars that passed through an intersection in 30 minutes, you could show the number of each color of car on a bar graph and compare the information.

Choose something to count in your classroom, around your home, or outside. It could be colors of cars passing through an intersection, the number of students wearing sneakers or boots, and so forth. Just choose something that is easy to count that has many subcategories. Create a bar graph with your information. Then determine the highest-occurring number and lowest-occurring number of items.

Election Results

Imagine that your class held an election to choose the treasurer for your next fund-raising event. The results of the election are shown in the chart below. Use the chart to answer the questions below.

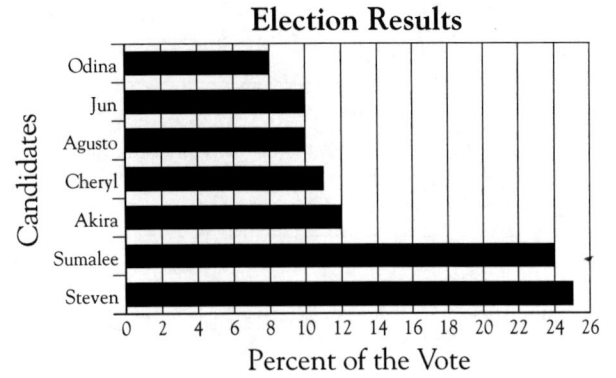

Election Results

Candidates (top to bottom): Odina, Jun, Agusto, Cheryl, Akira, Sumalee, Steven
Percent of the Vote

1. Who got the largest percent of the vote?
2. What two people received the same percent of the vote?
3. A majority would be more than 50%. Did anyone get a majority?
4. Who received the least number of votes?
5. Who was Steven's closest competition?

Daily Warm-Ups: Math in Real Life

113

© 2007 Walch Publishing

Reading Line Graphs

Line graphs are helpful if you are trying to show how a trend behaved over time. The graph below shows how the temperature changed over the course of one hot day in the summer.

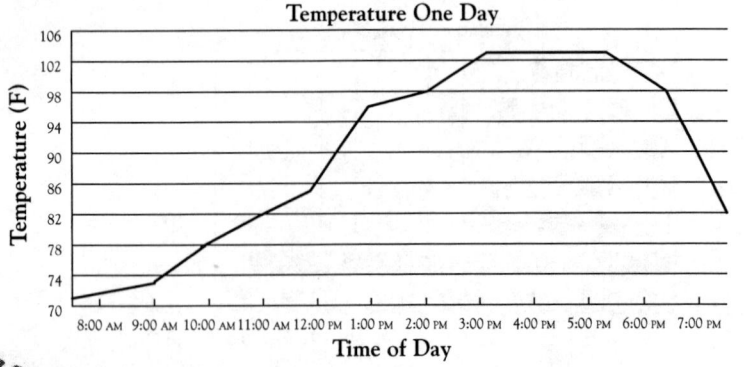

Temperature One Day

Answer the following questions about the graph above.

1. What was the lowest temperature between 8 A.M. and 7 P.M.?

2. What was the highest temperature between 8 A.M. and 7 P.M.?

3. Over what period did the temperature change the fastest?

4. When was the temperature 82 degrees?

5. What was the temperature at 6 P.M.?

Change Over Time

Line graphs are best used to show a change over time. They let us guess what was happening when we were not looking and can help us guess how something might have continued to behave after we couldn't see it any longer.

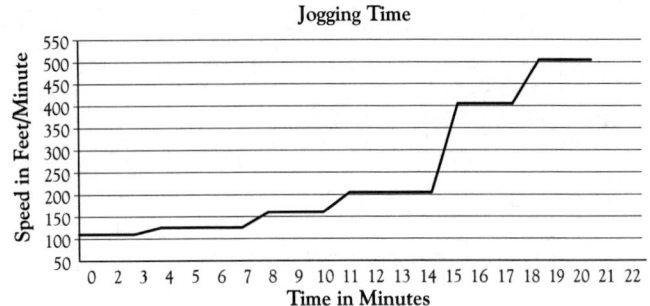

Answer the following questions about the graph above.

1. What is this graph describing?
2. What units are on the *x*-axis?
3. What units are on the *y*-axis?
4. When did the runner's speed increase the most?
5. What was the runner's highest speed?

Stock Graphs

The chart below shows the value of stock from the GE Company over a three-month period. The dates are on the x-axis and the value of the stock in dollars is on the y-axis. Use the chart to answer the questions that follow.

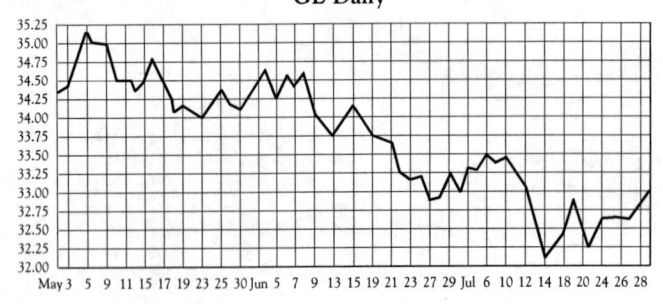

1. What was the maximum value of the GE stock during this period?

2. What was the minimum value of the GE stock during this period?

3. What would 100 shares have cost on June 13?

4. How much money would you have lost if you bought 1000 shares on May 9 and sold them on July 20?

5. How much money would you have made if you bought 5000 shares on July 18 and sold them on July 28?

Reading Scatter Plots

Below is a scatter plot showing the number of requests for food at a fast-food restaurant from 11 A.M. until 12:20 P.M.

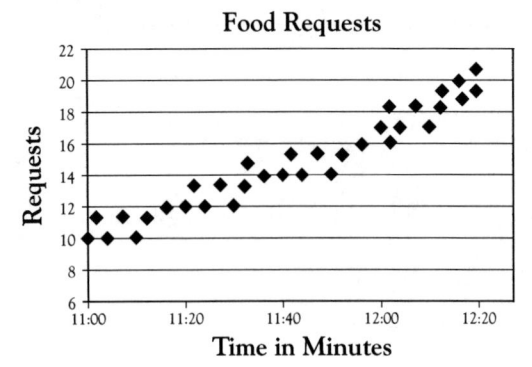

Food Requests

Answer the following about the scatter plot above.

1. When did the lowest number of requests per minute take place?

2. When did the highest number of requests per minute take place?

3. Why would the number of requests increase at a steady pace during this time of the day?

4. About how long did it take for the average number of requests per minute to go up from 10 to 16?

Histograms

A histogram is a way of combining a frequency table and a bar graph. The data is either collected or organized on a frequency table and then the results are transferred to a bar graph. The histogram below shows the grades earned on one class's final exam.

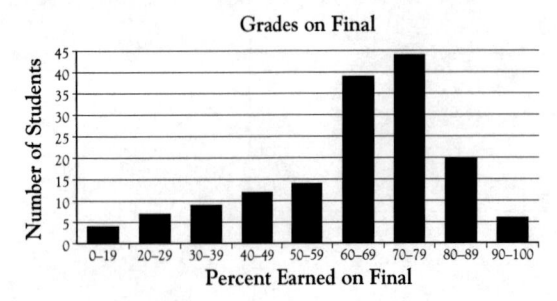

Answer the following questions about the histogram above.

1. How many students received a grade of 90 or higher?

2. What was the most common grade range?

3. What was the least common grade range?

4. A grade of 59 or lower is an *F*. About how many students failed?

5. About how many students passed?

Stem-and-Leaf Plots

A stem-and-leaf plot is a way of organizing a large collection of data. Once the data is collected, it becomes easier to see things such as which values are largest or smallest, or which value occurs most often.

Suppose your teacher grades on a curve and shows you the grades received by all the students in your class on the last test. The grades are as follows: 78, 100, 89, 45, 100, 94, 95, 95, 83, 83, 82, 74, 76, 79, 61, 90, 58, 65, 66, 89, 87, 91, 85, 76. Use this information to complete the following.

1. Put these grades into a stem-and-leaf plot.
2. Your teacher likes it when most of the grades are 80 or higher. Are they?
3. What category has the most leaves?
4. What category has the least leaves?
5. Your teacher likes the class average to be between 75 and 85. Find the class average.

High Jump

A person trying to get over the bar at the high jump is trying to build up a lot of kinetic energy (the energy of motion) and then convert that into gravitational potential energy (energy stored as an object is lifted against gravity). The best high jumpers are those who can most effectively convert their kinetic energy to potential energy. The formulas for kinetic and potential energy can be combined to show that $v = \sqrt{2gh}$, where v is the velocity (in meters/second), g is the acceleration due to gravity ($9.8 \frac{m}{s^2}$), and h is the person's height above the ground in meters.

Solve each problem below.

1. How fast should a person be running to get 2.2 meters above the ground?

2. What height can be achieved by a person running at 6.3 meters per second?

3. How fast should a person be running to set a world record height of 2.6 meters?

4. What height, theoretically, could a person running at 10 meters per second achieve?

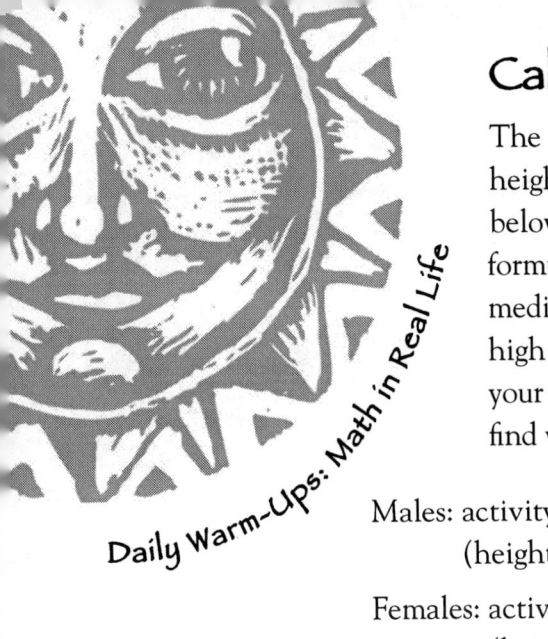

Calories per Day

The number of calories you should consume daily depends on your height, weight, age, gender, and activity level. Use the formulas below to calculate the number of calories needed daily. For each formula, a low activity level (little to no exercise) is worth 1.1, a medium activity level (some regular exercise) is worth 1.5, and a high activity level (training for competition) is worth 2.0. To find your weight in kilograms, divide your weight in pounds by 2.2. To find your height in centimeters, divide your height in inches by 2.54.

Males: activity level × [66 + (weight in kilograms × 13.7) + (height in cm × 5) − (age × 6.8)]

Females: activity level × [655 + (weight in kilograms × 9.6) + (height in cm × 1.85) − (age × 4.7)]

Complete the following.

1. Determine the number of calories you should eat in a day.

2. Determine the number of calories three of your classmates should eat in a day. Be sure to do the calculations for each gender.

Target Heart Rate

Your target heart rate is the rate that your heart should be beating while you are engaged in vigorous exercise. This allows your lungs and your heart to get the maximum benefit from your workout. Below are the formulas for calculating the target heart rate for both males and females at 85% of maximum heart rate. The formula is different for males and females because, in general, a woman's heart beats a little faster than a man's.

Target heart rate (male) = $0.85(220 - \text{age})$

Target heart rate (female) = $0.85(226 - \text{age})$

Complete the following.

1. Find your own target heart rate.
2. Find the target heart rate of someone of the opposite gender.
3. What is the target heart rate for a 65-year-old man? A 65-year-old woman?
4. What is the target heart rate for a 25-year-old man? A 25-year-old woman?

Body Mass Index

The body mass index (BMI) is a way of comparing your height and your weight. Having a high BMI can be very unhealthy. You can find your BMI by using the formula and chart below.

$$BMI = \frac{(\text{height in inches})^2}{(\text{weight in pounds} \times 703)}$$

BMI	Weight category
below 18.5	underweight
18.5 – 24.9	normal
25.0 – 29.9	overweight
30.0 and above	obese

Calculate the BMI for each person below.

1. a person who is 6 feet, 5 inches tall and weighs 240 pounds

2. a person who is 5 feet, 5 inches tall and weighs 140 pounds

3. a person who is 5 feet, 2 inches tall and weighs 100 pounds

4. a person who is 6 feet, 0 inches tall and weighs 400 pounds

Converting Temperatures

The United States still uses the Fahrenheit system for measuring temperatures, but most of the world uses the Celsius scale. On the Fahrenheit scale, 25° means chilly weather, but 25°C means a nice warm day. Below are the formulas for converting Fahrenheit temperatures into Celsius and Celsius temperatures into Fahrenheit.

Celsius to Fahrenheit: $°F = (\frac{9}{5} \cdot °C) + 32$

Fahrenheit to Celsius: $°C = (°F - 32) \cdot \frac{5}{9}$

Use the formulas above to convert the following temperatures.

1. 20°C

2. 30°C

3. 100°C

4. 75°F

5. −40°F

6. 100°F

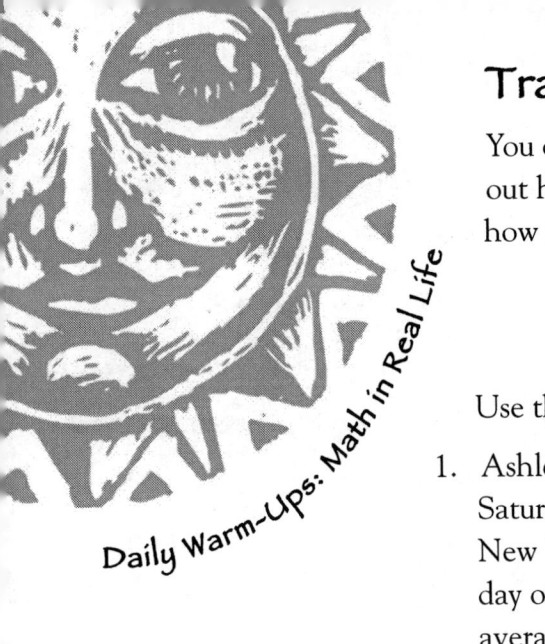

Travel Plans

You can use information about distance, rate, and time to figure out how long a trip will take, how far away a destination is, or how fast you are traveling.

$$\text{distance} = \text{rate} \times \text{time}$$

Use the formula for distance to complete the problems below.

1. Ashlee is driving from New York to Miami for a wedding on Saturday. It is 1298 miles from Ashlee's house to Miami. She leaves New York on Wednesday. If she drives no more than 8 hours a day on Wednesday, Thursday, and Friday, what will her average speed be?

2. Jorge is driving from Los Angeles to Sacramento. The trip is 368 miles, and he drives at an average rate of 62 miles per hour. How long will it take Jorge to get to Sacramento?

3. Sondra is driving home from college for the holidays. She drives at an average rate of 57 miles per hour, and it takes her 4.5 hours. How far is it from college to home?

Slugging Average

While watching a baseball game one evening, you hear the announcers talk about "slugging averages" and realize that you do not know how to calculate them. A quick search on the Internet shows that a slugging average is a measure of how far a batter got each time at bat. The formula is slugging average = $(s + 2d + 3t + 4h) \div AB$, where s is singles, d is doubles, t is triples, h is home runs, and AB is at bats.

Player	At bats	Singles	Doubles	Triples	Home runs	Batting average
Armas	427	60	21	1	41	.288
Masaru	404	76	24	0	30	.322
Brown	424	79	28	4	27	.325

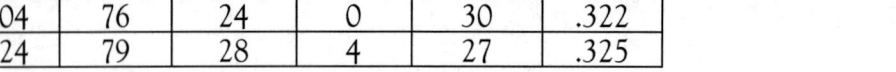

Complete the following using the chart above.

1. Calculate the slugging average of each of the three players.

2. Why might Armas and Masaru be considered better players than Brown?

Geometry Terms

There are a number of geometric shapes that you see every day. There are also a number of terms from geometry that you hear used in everyday speech. In the space below, make a list of all of the geometric shapes you can think of. Then make a second list of all the geometry-related terms that you can come up with. Here are two geometry-related terms to get you started: *altitude* and *rotation*.

Geometric Shapes in Nature

While walking through a park, a field, a forest, or a natural history museum, you can see a variety of geometric shapes. Animals are covered with patterns, rocks are made of crystals, and plants grow in hundreds of shapes and designs.

Identify the geometric shape of each item below.

1. a grain of salt

2. a single cell in a honeycomb

3. an eyeball

4. a carrot

5. a tree trunk

Human-Made Geometry

There are some geometric shapes that almost never occur in nature. For instance, there are few squares found in nature, and it's almost as difficult to find a triangle. These shapes do occur occasionally, but they are not as common as shapes such as hexagons or spheres. Humans, however, have found ways to make use of almost every kind of geometric shape.

Make a list of as many geometric shapes as you can think of. Then think of a human-made example of each of these shapes.

Magnifying Drawings

Being an artist can be a gratifying experience. However, what if you draw a picture that you want to enlarge? You could photocopy it and blow it up, but at that point you might be using technology more than art. If you want to magnify your drawing, you can use some graph paper to help you.

Complete the following.

1. On a sheet of graph paper, mark off a section that is 8 squares by 8 squares.

2. Draw a simple picture or geometric pattern in the area.

3. Mark off a new section that is 16 squares by 16 squares.

4. Copy the image of each square of your original picture into a new section that is 2 squares by 2 squares.

5. How much larger is your new picture than your old one?

Symmetry

A line of symmetry is a line that can be drawn through a shape so that it cuts the figure into two pieces that are mirror images of each other. Symmetry has been used to construct buildings and to create works of art for thousands of years.

List ten everyday items that are symmetrical. For each item, indicate where the line of symmetry is or explain if it has more than one line of symmetry.

Perimeter

Perimeter is the distance around a closed shape. The perimeter of a shape can be found by adding the lengths of all the individual sides.

Solve each problem below.

1. You have to buy some fencing to go around your yard. The distance across the front is 100 feet, down each side is 250 feet, and across the back is 120 feet. How much fencing do you need?

2. You need to buy trim to go around a tablecloth that is 8 feet long and 3 feet wide. How much trim will you need?

3. Your new house has an unfinished room over the garage that is 12 feet by 24 feet. There is a floor, but no baseboards. The door and its frame are 4 feet wide. How many feet of baseboard will you have to buy?

4. You have to tape a 12-foot by 17-foot room before you paint it. There are three windows that are 4 feet by 6 feet, two doors that are 3 feet by 7 feet, and baseboards that circle the room. How many feet of tape will you need?

Daily Warm-Ups: Math in Real Life

Circumference of a Circle

Circles are shapes that you see every day. The point at the middle of a circle is called the center. The distance across the circle that passes through the center is called the diameter, and half of the diameter is the radius. The distance around the outside of a circle is called the circumference, and the ratio of the circumference to the diameter of a circle is called *pi*. *Pi* is usually abbreviated as 3.14. The circumference of a circle is found using the formulas $C = \pi d$ or $C = 2\pi r$, where d is diameter, r is the radius, and π is *pi*.

Solve each problem below.

1. You need to buy ribbon to decorate two columns in front of your home. Each column has a radius of 2 feet. If you want to go around each column 10 times, how much ribbon will you need?

2. You are designing a new label for a soda company. The radius of a soda can is about 3.2 centimeters. How long will the label have to be to go entirely around the can?

Area of a Square

Squares are shapes you see every day. The area of a square can be found by multiplying the length of two adjacent sides. In a square, all the sides are the same length. If you know the length of any one side, you can square it to find the area.

Solve each problem below.

1. A cat climbs the screen in a screen door and tears a hole in it. You buy some extra-strength screening to replace it. The square screen is 90 centimeters on each side. How many square centimeters of screening do you need?

2. You want to put up a square canvas to shade your circular above-ground swimming pool from the sun. If the radius of the pool is 5 feet, how many square feet of canvas do you need?

3. Square poster board is 3 feet by 3 feet. How many square feet of poster board are in a 12-pack?

Area of a Rectangle

Rectangles are shapes you see every day in many different places. A rectangle must have four right (90°) angles at the corners and have opposite sides that are the same length. The formula for the area of a rectangle is area equals length times width, or $A = l \cdot w$.

Solve each problem below.

1. You decide to buy an area rug for your dining room that is 4 feet by 8 feet. What is the area of the rug?

2. An official NCAA football field is 140 feet wide and 360 feet long including the end zones. How many square feet is the entire field?

3. The runway for the space shuttle is 15,000 feet long and 300 feet wide. What is the area of the runway?

4. How many NCAA football fields could fit onto the runway for the space shuttle?

Area of a Circle

Remember that the formula for calculating the area of a circle is $A = \pi r^2$. The A stands for area, the r stands for radius, and the π is equal to 3.14.

Solve each problem below.

1. Your beach umbrella is 6 feet wide. How many square feet of shade will it provide?

2. You order a 12-inch pizza and an 18-inch pizza. Which is more, half the 18-inch pizza or $\frac{3}{4}$ of the 12-inch pizza?

3. You have a stack of soda cans that is 20 cans by 20 cans at its base. How much surface area is actually covered by can bottoms if the radius of a soda can is about 3.2 centimeters?

Sprinklers

Your lawn is a square that is 60 feet by 60 feet. You have a sprinkler that has a diameter of 30 feet. In the space below, create a drawing that shows one possible way you could water the entire lawn while moving the sprinkler as few times as possible. Draw an *x* in each place where you are going to put the sprinkler. (*Hint*: It can be done in as few as 13 moves, maybe less!)

Area of a Triangle

Triangles are shapes you see every day in many different places. The formula for the area of a triangle is $A = \frac{1}{2}bh$, where A is the area, b is the length of the base, and h is the height of the triangle.

Solve each problem below.

1. Your square lawn is 100 feet long on a side. You and your brother each agree to mow half of the lawn so you split it down the middle diagonally. How many square feet of the lawn will each of you mow?

2. A tortilla chip is 7 centimeters on a side and has a height of 5.5 centimeters. How much area is there on a tortilla chip for salsa?

3. You buy a yield sign from a traffic sign supplier to hang on your bedroom door. The sign is 30 inches on a side and 15 inches in height. How many square inches of your door does it cover?

Subtracting Areas

It would be simple if every measurement you made was perfectly even or if every area was a perfect rectangle, but that is seldom the case. Often you will find that something you are trying to measure has an irregular shape or that some other object interferes with a simple measurement.

Solve each problem below.

1. You plan to carpet a room that is 12 feet by 16 feet. You will be putting ceramic tile in front of the fireplace in a spot that is 2 feet by 3 feet. How many square feet of carpeting do you need?

2. You need new sod for your lawn, which is 50 feet by 50 feet. You have an above-ground swimming pool with a diameter of 12 feet in the middle of your lawn. How much sod will you need? (Remember the formula for finding the area of a circle: area = πr^2)

Area of a Compound Figure

A compound figure is made up of other more common shapes. To find the area of a compound figure, you can break the figure into common shapes and calculate the areas of those common shapes. When you have the individual areas, you can add them together to find the total area of the compound figure.

Solve each problem below. Round to the nearest whole unit. (Remember the formula for finding the area of a circle: area = πr^2)

1. Your new office desk is made of up two rectangles that are each 24 inches by 36 inches and a centerpiece that is one quarter of a circle with a radius of 24 inches. What is the total surface area of the desk?

2. Your dining room table contains two leaves that are 36 inches by 12 inches each. When they are removed, the remainder of the table is a perfect circle with a diameter of 36 inches. What is the total area of the table when both of the leaves are in?

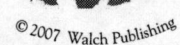

Surface Area of a Cube

If you know the length of any one side of a cube, you can calculate the surface area of all the sides. Just square the length to find the area of one side and then multiply the answer by 6. There are 6 sides on a cube, and they all have the same area.

Solve each problem below.

1. A Rubik's Cube is 2.25 inches on a side. What is the surface area of the entire cube?

2. A grain of salt is about 70 micrometers on a side. If you and a lab partner measure a grain of salt and find that the grain is 64 micrometers on a side, what is the surface area of the grain of salt?

3. You are making a footstool and want to cover it in leather. The cube will be covered on all sides, and each side is 18.5 inches. How many square feet of material will it take to cover all six sides?

Surface Area of an Oblong Box

If you take an oblong box such as a tissue box and unfold it, you can see that the box is made up of six rectangles, and opposite rectangles are equal in size. The formula for the surface area of a box is $S = 2(lw + hw + lh)$.

Solve each problem below.

1. You are helping a friend install soundproofing in his new recording studio. The main recording room will be soundproofed on all four walls, the floor, and the ceiling. The room is 24 feet long, 12 feet wide, and 8 feet tall. How many square feet of soundproofing material will you need?

2. You are packing a car to be shipped overseas and need lumber to make a crate to ship it in. The car is 181 inches long, 66 inches tall, and 72 inches wide. How many square feet of lumber will it take to make a box for the car?

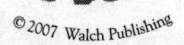

Cleaning Crew

You get a job with a company that cleans windows. The company sends crews to a number of buildings in a large city. Some buildings are so large that by the time the crew cleans all the windows, the ones they cleaned first are dirty again.

Solve each problem below.

1. One of the buildings is 800 feet tall and has a base that is 200 feet on a side. What is the amount of surface area of the windows if they completely cover all four sides?

2. The largest building is 840 feet tall and has a base that is 175 feet by 245 feet. The surface of the building is covered with windows that are 9 feet by 6 feet with 1-foot frames in between. This means that the building is 84 windows tall, 25 windows down its short side, and 35 windows down its long side.

    a. What is the surface area of the entire building?

    b. What is the surface area of just the windows?

    c. What is the surface area of all the frames?

Surface Area of a Sphere

There are spheres all around us. Marbles, bowling balls, and tennis balls are kinds of spheres. To find the surface area of a sphere, you can use the formula $S = 4\pi r^2$. The r in the formula stands for the radius of a sphere, which is the distance from the center of the sphere out to a point on the surface.

Solve each problem below.

1. Your job at a basketball factory is ordering material for making basketballs. A standard men's basketball is 9.52 inches in diameter. If you are going to make 1100 basketballs in one day, how many square feet of material will you need?

2. A standard women's basketball is 9.15 inches in diameter. How much less surface area does each women's basketball have than each men's basketball?

3. You get a job painting a protective coating on expensive handmade bowling balls. The can says you have enough to paint 400 square feet of surface. If the bowling balls each have a radius of 4.3 inches, how many should you be able to paint with one can?

Surface Area of a Pyramid

There is no simple formula for the surface area of a pyramid because the shape of the base could be any one of a number of different shapes. The total surface area is found by carefully counting all the faces and then calculating the surface area for each one. When this is done, you must add all the individual surface areas to find the total surface area for the whole pyramid.

Solve each problem below. (Remember, surface area = $\frac{1}{2}bh$.)

1. Most of the original casing stones of the Great Pyramid of Giza in Egypt were loosened by an earthquake in 1301. The base of the pyramid is a square that is 215.3 meters on a side. The distance up the middle of any of the four triangular faces of the pyramid is 179.7 meters. How many square meters of casing stone would resurface the pyramid?

2. The Louvre Museum in France has a glass pyramid that is about 35.42 meters long on each side of its square base. The distance up the center of each triangular face is about 29.6 meters. How many square meters of glass make up the pyramid?

Surface Area of a Cylinder

A cylinder is an object with two sides that are parallel and with congruent surfaces that are both circles. A soda can is an example of a cylinder. The formula for the surface area of a cylinder is $S = 2\pi r^2 + 2\pi rh$, where h is the height of the cylinder.

Solve each problem below.

1. The two stone columns in front of your office building need to be decorated with silver foil for the holidays. Each column is 30 feet tall and has a radius of 1.5 feet. How much foil will be needed?

2. You are in charge of developing a new product at a canning company. The can must be large enough for a family-size serving of fruit cocktail. This means that the can must be 111 millimeters tall and 100 millimeters in diameter. How many square millimeters of material will be needed to make each can?

Volume of a Sphere

Bowling balls, racquetballs, marbles, gumballs, and many other objects are spheres. To calculate the volume of a sphere, you need to know the radius of the sphere and use the formula $V = \frac{4}{3}\pi r^3$. In the formula, r is the radius. Be sure to remember that the radius is cubed, not squared.

Solve each problem below.

1. A basketball has a diameter of 9 inches. What is its volume?

2. How many basketballs will fit into a box that is 2 feet by 2 feet by 3 feet?

3. A tennis ball has a diameter of 2.5 inches. What is its volume?

4. How many tennis balls will fit inside the same box?

Volume of a Cube

A cube is also known as a regular hexahedron. A cube has six faces that are made of sides with equal lengths and angles with equal measurements. To find the volume of a cube, you only need to know the length of any one side. The formula for the volume of a cube is $V = l^3$, where l is the length of any one side.

Solve each problem below.

1. You have a storage facility that is made up entirely of cubical rooms that are 8 feet long, 8 feet wide, and 8 feet high. If the facility has 320 rooms, how many cubic feet of storage are you offering to your customers?

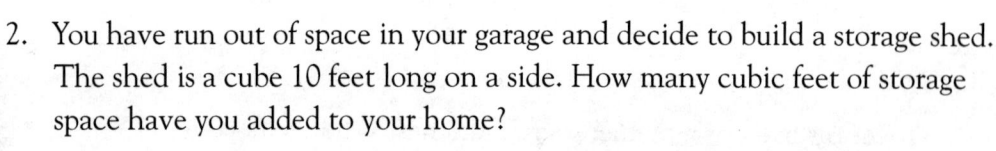

2. You have run out of space in your garage and decide to build a storage shed. The shed is a cube 10 feet long on a side. How many cubic feet of storage space have you added to your home?

3. A decorative box of tissues is a cube 4 inches on a side. If a tissue takes up 0.533 cubic inches, then how many should fit in a single box?

Volume of a Cone

A cone is an object that has two sides—a circular base and a slanted face. The cone is a little unusual because it doesn't have edges or a vertex. The basic formula for the volume of a cone is $V = \frac{1}{3}bh$. In this case, the area represented by the variable b is the area of a circle, or $A = \pi r^2$. If we use πr^2 in place of the variable b, the formula becomes $V = \frac{1}{3}\pi r^2 h$.

Solve each problem below.

1. Your boss at the ice cream store has told you to stop filling up the sugar cones with ice cream. How many cubic inches of ice cream are the customers losing out on if the cone has a diameter of 2 inches and a height of 5.5 inches?

2. You are putting up a spotlight to light a dark corner of your room. The light fixture makes a cone of light that has a diameter of 4 feet and is 8 feet high. What volume of the room is taken up by the light?

Volume of an Oblong Box

An oblong box has length, width, and height. By multiplying the three together, you can find the volume. The formula for the volume of an oblong box is $V = l \cdot w \cdot h$. The units for volume are cubes, as in cubic centimeters or cubic inches.

Solve each problem below.

1. The closet in your bedroom is 6 feet wide, 3 feet deep, and 8 feet high. How many cubic feet of storage do you have?

2. The interior dimensions of a small train boxcar are 50 feet long, 9 feet wide, and 10 feet high. The dimensions of a large boxcar are 60 feet long, 9 feet wide, and 13 feet high. You have to move 300,000 cubic feet of lumber. How many more of the small boxcars would it take than large boxcars?

Volume of a Cylinder

A cylinder is an object with two sides that are parallel and with congruent surfaces that are both circles. The volume of a cylinder can be found by multiplying the height of the cylinder by the area of the circle on the top or bottom of the cylinder. The formula for the volume of a cylinder is $V = \pi r^2 h$, where V is the volume, r is the radius of the circle on the top or bottom of the cylinder, and h is the height of the cylinder.

Solve each problem below.

1. You measure a 1-gallon paint can and find that it has interior measures of diameter 6.482 inches and height 7 inches. How many cubic inches are in a gallon?

2. Barrels of oil have 9702 cubic inches of oil in them. Based on your answer to number 1, about how many gallons are in a barrel of oil?

Volume of a Pyramid

A pyramid is a three-dimensional shape that has at least three faces that are triangles and only one base. To calculate the volume of a pyramid, you have to calculate the volume of the base. The height of the pyramid is a line that runs from the peak of the pyramid down to the base so that it forms a right angle with the base. The formula for the volume of a pyramid is $V = \frac{1}{3}bh$, where b stands for the area of the base and h stands for the height of the pyramid.

Solve each problem below.

1. You buy a new house and discover that it has an uninsulated attic. The house has a pyramid roof (or hip roof), and the interior dimensions are 36 feet by 36 feet. The ceiling is 6 feet high at its center. If you want to fill the whole space with insulating foam, how many cubic feet will you need?

2. The Great Pyramid of Giza has a square base that is 215.3 meters on a side. It has a vertical height of 143.9 meters. A cubic meter is 35.5 cubic feet. How many cubic feet are there in the pyramid?

Right Triangles

A right triangle has one angle that has a measure of 90°. The side of the triangle that is opposite the 90° angle is called the hypotenuse. The other two sides of the triangle are called the legs. The Pythagorean theorem allows you to find the missing length of any side of a right triangle if you know the lengths of the other two sides. The formula is $a^2 + b^2 = c^2$, where a and b are the lengths of the two legs and c is the length of the hypotenuse.

Solve each problem below.

1. You're building a ramp, but the distance is so long that your tape measure can't get an accurate reading. The top of the ramp will be 6 feet above the ground, and the horizontal distance of the ramp is 60 feet. How long will the diagonal of the ramp be?

2. While building a roof on your garage, you find that for every 4 feet of horizontal distance, your vertical distance has to increase 3 feet. How long will the slanted part of the roof be for every 4-foot horizontal run?

Survey

Advertisers like to know how well they are matching their products to the consumers who are likely to buy them. By polling and surveying groups of people, advertisers can find out how effective their ads are. When a survey is conducted, a respondent can give any answer he or she chooses instead of selecting from a list of possible responses.

Take a survey of your classmates to find each person's favorite color. Draw a bar graph to show the results. Then answer the following questions.

1. What was the most common response?

2. What was the least common response?

The Survey Says...

A local radio station is doing a survey at a nearby mall. You and your friends agree to answer some questions about your favorite music to get your choice of a free CD or some music downloads. Two weeks later, the station sends you a copy of the results of their survey. According to the survey, 18% liked country music, 22% liked rap and hip-hop, 30% liked pop/rock, 9% liked classical, and the remaining 21% liked jazz, blues, gospel, bluegrass, dance music, oldies, or classic rock.

Answer the following questions.

1. Why would a radio station be interested in the types of music that people enjoy?

2. What type of music was liked by the most people?

3. Is the mall the best place for this type of survey? Why or why not?

4. Based on the results of the survey, what age group spends a lot of time at the mall?

Playing Favorites

Create a survey that asks just one question: What is your favorite television show? Then follow the directions below.

1. Collect data from all the female students in your class.
2. Collect the same data from all the male students in your class.
3. Make two bar graphs, one from each group, that show the results of your survey.
4. What differences do you see? How might this information be useful to advertisers?

Multiple Choice

When completing a questionnaire, a respondent might select from a list of possible responses to prevent the data collection from becoming too complicated. For instance, 100 different people might give 100 different answers when asked to name their favorite movie. If respondents are given a list of four choices, however, they can easily choose the movie they like the most.

Imagine that you have a job at a local deli making sandwiches. Design a questionnaire to help your boss determine the various breads, meats, cheeses, and so forth that local customers would enjoy the most. Make sure you give respondents possible responses to choose from for each question.

Independent Events

Independent events are not connected to one another. For example, the odds of you getting hit by a runaway camel and being hit by lightning on the same day are incredibly small. But the two events are not really connected in any meaningful way. To find the probability of two independent events, you can multiply the odds of the first event by the odds of the second event.

Solve each problem below.

1. You buy a raffle ticket for a fund-raiser. You find that the drawing will be from 345 tickets. What are the odds of you winning?

2. You flip a quarter 3 times in a row. What are the odds of getting heads all 3 times?

3. What are the odds of you winning the raffle from number 1 and getting heads 3 times in a row from flipping a quarter?

Dependent Events

Dependent events are occurrences in which the outcome of one event affects the outcome of another event. For example, if you were to put 4 red marbles and 4 blue marbles in a bag, the odds of pulling out a red marble would be $\frac{4}{8}$. The odds of pulling out a blue marble next would be $\frac{4}{7}$. The total odds would be $\frac{4}{8} \times \frac{4}{7} = \frac{16}{56}$, which reduces to $\frac{2}{7}$ or about 28.6%.

Solve each problem below.

1. Your desk drawer contains 12 blue pens, 12 black pens, and 10 pencils. What are the odds of pulling out 1 blue pen, 1 black pen, and 1 pencil?

2. What are the odds of pulling out all 12 blue pens?

3. What are the odds of pulling out 3 pens and 3 pencils?

Control Group

A *control group* is a group that remains under "normal" conditions during an experiment for comparison. Say you grow sunflowers every year. If you wanted to find out if moving their location would make them bigger, you would plant them as you usually do and then plant a second group in a new location. You would keep everything else the same. Then you could compare the group in the new location with the control group in the old location.

Complete the following.

1. Your teacher says she has a new way to teach a difficult section of math. She has two classes that perform similarly. She chooses your class (Class 1) to test the new teaching method on. She teaches the "old" way to the other class. Find the quiz average for each class after using the new method.

    Class 1: 81, 92, 67, 43, 99, 99, 96, 88, 90, 70, 75, 88, 79, 86

    Class 2: 76, 87, 62, 38, 94, 94, 91, 83, 85, 65, 70, 83, 74, 81

2. Does the new teaching method seem effective or not?

Permutations

A *permutation* is a set of outcomes that appear as an arrangement or a list where the order matters. For example, if you were deciding the order that 3 friends could use your new video game system, you would see that the combinations for friends A, B, and C are:

ABC, ACB, BAC, BCA, CAB, and CBA. The number of permutations are $3 \times 2 \times 1 = 6$.

Solve each problem below.

1. You get a big-screen television and decide to have 5 friends over to watch a movie. They always disagree about who gets to sit where, so you decide to make a seating chart before they get there. How many permutations are there for the seating arrangement?

2. You are keeping score at a cross-country race at your school and will be writing down the order that the runners finish. There are only 14 people in the race, so how many permutations are there for the first, second, and third place finishers?

Speaking Up

You organize a presentation for a local charity and have 5 speakers who volunteer to teach your team how to raise more money. There is no particular order that the speakers have to go in.

Answer the following questions.

1. How many different speakers could go first?

2. Once the first speaker goes, how many speakers are left to go second?

3. Once two speakers have gone, how many speakers are left to go third?

4. Once three speakers have gone, how many speakers are left to go fourth?

5. Once four speakers have gone, how many speakers are left to go fifth?

6. How many different orders could the speakers have given their speeches in?

Decisions, Decisions

When you have to make a decision, the more factors there are to consider, the harder the decision can be. Sometimes the best way to make a decision is to know all the possible choices so that you can eliminate them.

Solve each problem below.

1. You have 3 kinds of bread, 3 kinds of cheese, 3 kinds of meat, and 6 kinds of vegetables. How many different kinds of sandwiches could you make with these ingredients, assuming that each sandwich only has 1 kind of bread, 1 kind of cheese, 1 kind of meat, and 1 kind of vegetable?

2. Tired of making sandwiches, you go out to eat. There are 12 appetizers, 25 entrées, 35 drinks, and 16 desserts on the menu. How many different meals could you have assuming you got 1 appetizer, 1 entrée, 1 drink, and 1 dessert?

Combinations

A *combination* is a list of possible groupings in which the order does not matter. For example, if you wanted to get pizzas that had all the combinations of mushrooms, onions, and peppers, taken two at a time, the combinations are as follows:

- a. mushrooms, onions
- b. mushrooms, peppers
- c. onions, peppers
- d. onions, mushrooms
- e. peppers, mushrooms
- f. peppers, onions

The last three combinations are repeats of earlier combinations. The expression to find 3 things taken 2 at a time is $\frac{3!}{2!(3-2)!}$.

Complete the following.

1. You go out to dinner and see that the dessert menu has 6 kinds of ice cream. You can get 3 scoops of any combination of flavors you want. How many combinations are there?

2. You and a group of friends join a basketball team. There are 12 people. How many combinations of 5 could be on the floor at different times?

Soccer Tournament

Your school is hosting a soccer tournament that will have 8 teams involved. You are working with the athletic director to set up a game schedule.

Solve each problem below using the following formula:

$$\frac{T!}{S!(T-S)!} \quad \begin{array}{l} T\text{—total} \\ S\text{—selected set} \end{array}$$

1. How many possible ways are there to match up the teams for the games in the first round?

2. In the second round, there will only be 4 teams left. How many ways are there for the 4 teams to match up?

3. How many possible different games could there be from the beginning to the end of the tournament?

4. What is the mathematical probability (ignoring the team records) of any one team winning?

Password Protected

As more and more information is stored on computers, people are at greater risk of having their personal information stolen, being a victim of identity theft, and having their computer systems compromised. One of the most effective ways to prevent such problems is to have complicated password protection on your computer.

Solve each problem below using the following formula:

$$\frac{T!}{S!(T-S)!}$$

T—total
S—selected set

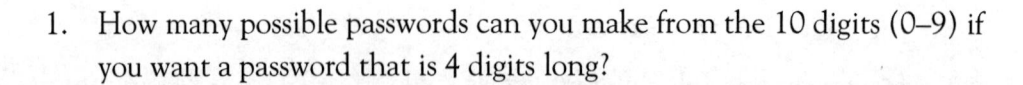

1. How many possible passwords can you make from the 10 digits (0–9) if you want a password that is 4 digits long?

2. How many possible passwords can you make from the 10 digits (0–9) if you want a password that is 5 digits long?

3. If you want to increase security by making more password options possible, how could you do that?

Word Jumbles

You may be familiar with word jumbles from seeing them in the newspaper or in puzzle books. In a word jumble, a relatively short word has its letters scrambled, and you have to unscramble them to see what the word is.

For each of the following, calculate the number of possible combinations that the set of letters could have. Then determine what the word is.

1. amny
2. horst
3. bejlum
4. cesiton
5. amdering
6. abmudshot

Relative Frequency

You have probably noticed that the letters on a computer keyboard are not in alphabetical order. This is also true of many portable electronic devices that you can use to send e-mail or text messages. There is a good reason for the arrangement of such keyboards.

Complete the following.

1. Find a standard fiction book and count all the occurrences of each letter on a page. Make a list from *a* to *z*, and then make a tick mark every time you come to one of the letters.

2. Count the total number of tick marks for each letter.

3. Determine the total number of letters you counted.

4. Calculate the percentage that each letter makes up of the total.

5. Why do you think the letters on a computer keyboard are grouped the way they are?

Mean

The *mean* is the sum of all the numbers in a group divided by the number of numbers in the group. It is the same as the average. For example, the group of numbers 18, 23, 12, 54, 39, 58, and 27 consists of seven numbers. The sum of those numbers is 231. 231 ÷ 7 = 33. So the mean or the average is 33.

Complete the following.

1. Collect the ages of all the people in your class and find the mean.

2. Collect the heights of all the people in your class and find the mean.

3. The mean weight of three people is 145 pounds. If one person weighs 105 pounds and one person weighs 205 pounds, what does the third person weigh?

Median

The *median* for a group of numbers is the middle number when the group has been ordered from smallest to largest. If there is no central number, then you find the mean of the two middle numbers. You might want to find the median of a group of numbers to organize them when there are many numbers close together and some numbers are large or small compared to the rest of the group.

Complete the following.

1. What is the median height of all the people in your class?

2. What is the median age of all the people in your class?

3. You and a group of friends go on a fishing trip and catch a number of fish of all different kinds. The lengths of the fish in inches are as follows: 10, 21, 16, 39, 28, 12, 50, 35, 35, 46, 23, 72, 18, 26, 29, 31, 28, 43, 41, and 56. What is the median length?

Mode

The *mode* is the number or numbers that occur most often in a group of numbers. You might want to find the mode of a group when there are many numbers that have the same value. This allows you to best understand which number out of many is most common.

Imagine that you have a job keeping track of the inventory in an electronics store. Your boss wants you to sort through a bin of electrical cords that contains 3-foot, 6-foot, and 9-foot cords. She asks you to determine which kind of cord you have the most of. As you sort through the cords, you find the following lengths:

3, 6, 6, 9, 3, 3, 6, 9, 9, 9, 9, 9, 9, 9, 6, 6, 3, 3, 6, 3, 9, 9, 6, 6, 3, 3, 3, 6, 3, 6, 3, 3, 6, 6, 6, 3, 6, 3, 9, 9, 6, 6, 9, 6, 9, 6, 9, 3, 6, 9, 9, 6, 6, 9, 3, 3, 6, 6, 3, 3, 3

Which length of cord do you have the most of?

Sports Means

You have two friends on the basketball team who have been given all the raw stats data for the season they have just finished. They are in disagreement as to which one is the better player. Emilio's stats are 14 games, 219 points, 39 rebounds, and 21 assists. Doug's stats are 18 games, 235 points, 46 rebounds, and 26 assists.

Complete the following.

1. Calculate the average number of points per game for each player.

2. Calculate the average number of rebounds per game for each player.

3. Calculate the average number of assists per game for each player.

4. Did either player perform consistently better on a game-per-game basis? If so, who?

Flipping Coins

You have a friend who wants to play a game with you. He has a stack of quarters, and he says, "Every time I flip a quarter and get heads 3 times in a row, you give me 4 quarters. Every time I get tails before I get 3 heads in a row, I give you 1 quarter."

Complete the following.

1. What are the odds that your friend can flip a quarter and get heads 3 times in a row?

2. If he succeeds and you pay him the 4 quarters, are you likely to get your money back if you keep playing?

3. Is this game to his advantage? Why or why not?

Rolling Dice

Many board games come with dice. The longer you play a game, the more it seems to matter what the outcome of rolling those dice will be. Each die has 6 sides, which means there is a 1 in 6 chance of any single number coming up with each roll.

Solve each problem below.

1. In one game, you like to get doubles because it allows you to roll again. What are the odds of getting doubles when rolling a pair of dice?

2. a. The goal of another game is to get five of a kind in three rolls. On the first roll, you roll five dice and one 6 comes up, which you set aside. On the second roll, you roll the remaining four dice and one 6 comes up, which you set aside. What are the odds that the last three dice will all be 6s on the third roll?

   b. What are the odds of getting five 5s on the third roll if you did not get any 5s on the first two rolls?

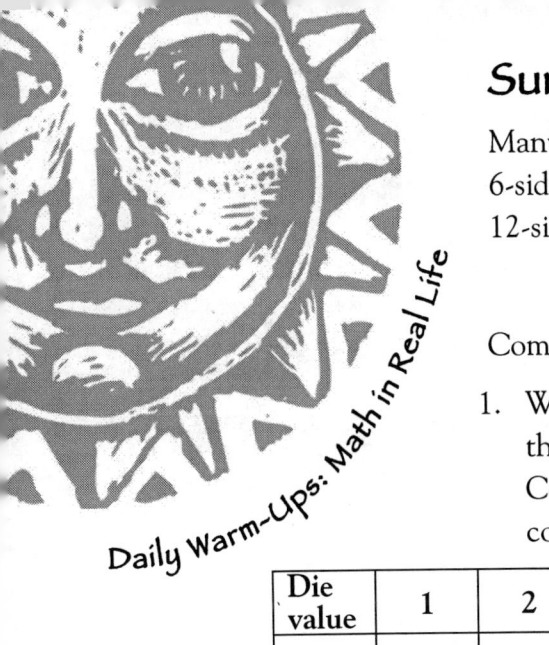

Sum of Two Dice

Many games use dice with more or fewer sides than a regular 6-sided die. For instance, there are 4-sided, 8-sided, 10-sided, 12-sided, and 20-sided dice, as well as a variety of others.

Complete the following.

1. While playing a game with a friend, you have to make a roll that will be the sum of the 4-sided die and the 6-sided die. Complete the chart below to show all the possible sums you could get.

Die value	1	2	3	4	5	6
1						
2						
3						
4						

2. Calculate the odds of getting each of the possible sums that you found.

Scrabble

The game of Scrabble starts with a bag of 98 tiles of letters in the alphabet and 2 blank tiles, making a total of 100 tiles. At the beginning of the game, tiles are drawn to see who goes first, and then the tiles are returned to the bag so that the first player can draw his or her 7 tiles from the entire range of 100. Below is a list of all the letters and how many of each one appears in the game.

A—9	E—12	I—9	M—2	Q—1	U—4	Y—2	
B—2	F—2	J—1	N—6	R—6	V—2	Z—1	
C—2	G—3	K—1	O—8	S—4	W—2	Blank—2	
D—4	H—2	L—4	P—2	T—6	X—1		

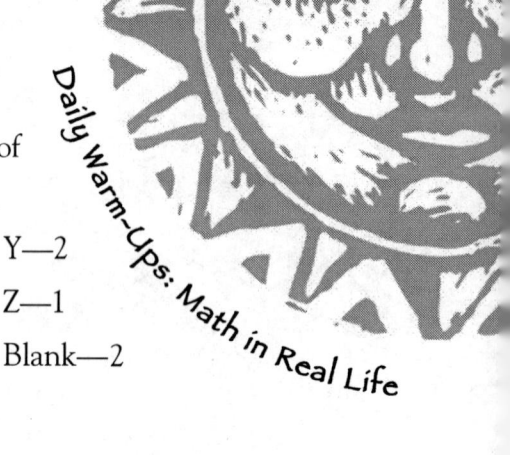

A perfect draw would be one in which you draw 7 letters that you could use all in one word. What are the odds of drawing the letters to spell each word listed below?

1. absolve _____

2. mansion _____

3. forgive _____

What Are the Odds?

The most common diseases that affect human beings have been widely researched. For many diseases, we know the odds of a man or a woman getting a certain disease. The odds of a woman under the age of 85 getting breast cancer are about 1 in 9. The odds of a man getting lung cancer if he smokes are about 78 per 100,000. The odds of a woman who smokes getting lung cancer are about 51 per 100,000. A number of factors increase the cancer rate, such as the age the smoker started smoking and the amount smoked.

Solve each problem below.

1. If there are 25,000 women in your town under the age of 85, then how many of them may develop breast cancer?

2. In 2006, there were approximately 150,100,000 women living in the United States. If 19.2% of these women smoked, how many cases of lung cancer could we expect among this population of women?

3. In 2006, there were approximately 145,600,000 men living in the United States. If 24.1% of these men smoked, how many cases of lung cancer could we expect among this population of men?

Lotto Luck

It is difficult to walk through a convenience store without seeing scratch lottery tickets or advertisements for state or national lotteries. The odds of winning money playing these games are usually very low.

Solve each problem below.

1. Your state lottery is offering scratch tickets for $1 each. The odds of winning $1 are 1:10, and the odds of winning $10 are 1:120. If you buy 5 tickets, what are your odds of winning $1? What are your odds of winning $10?

2. You are about to buy your first ticket for a national lottery. You have to match 5 numbers out of 45 and then one "magic ball" out of 50. What are your odds of winning if you buy 1 ticket? What are your odds of winning if you buy 100 tickets?

Daily Warm-Ups: Math in Real Life

178

© 2007 Walch Publishing

Playing Card Guess

A standard deck of playing cards has 52 cards in it. Of course, in almost every card game, getting the right card at the right time can help guarantee your win. There are 13 cards in each of the four suits: clubs, diamonds, hearts, and spades. The cards in the basic deck are the ace, 2, 3, 4, 5, 6, 7, 8, 9, 10, jack, queen, and king.

Solve each problem below.

1. What are the odds of drawing a queen from a full deck?

2. What are the odds of drawing two aces from a full deck?

Monday Birthday

To the nearest percent, the probability that any one person selected at random was born on a Monday is 14 percent (1 out of 7). What is the probability, to the nearest percent, that of any seven persons chosen at random, exactly one was born on a Monday?

Answer the question above using the following formula:

> The binomial distribution says that the probability of getting m hits out of n trials where the probability of a hit is p is given by:
>
> $$(n \text{ choose } m) * p^m * (1 - p)^{(n - m)}$$

Answer Key

1. 1. $0.28
    2. $0.80
    3. $0.78
    4. $0.41
    5. $0.74
    6. $0.67
2. e
3. 1. $105
    2. $380
    3. $85
    4. $866
    5. $759
    6. $188
4. c
5. 1. $270.36
    2. $279.98
    3. $403.11
    4. $316.40
    5. $371.25
    6. $145.80
6. c
7. 1. 5¢
    2. 15¢
    3. 1¢
    4. 6¢
    5. 6¢
8. 1. incorrect
    2. incorrect
    3. correct
    4. incorrect
    5. correct
9. 1. $4.74
    2. $37.66
    3. $127.05
    4. $15.60
    5. $22.68
    6. $33.66
10. 1. the six-pack
    2. same
    3. the 12-pack
    4. the 500-sheet ream
    5. the 50-pack
    6. each track for 75¢
11. Seat 1: $12, $12.29   Seat 3: $14, $14.46
    Seat 2: $10, $9.09
12. 1. discount = $1.88; sale price = $10.62
    2. discount = $9.19; sale price = $36.76
    3. discount = $187.50; sale price = $187.50
    4. discount = $435; sale price = $2465
13. 1. $87.50
    2. $630
    3. $22.7
    4. $398.40
    5. $336
    6. $75
14. 1. $119
    2. $1832
    3. $19.75
    4. $83.94
    5. $26.60
    6. $9429
15. 1. $11.07
    2. $18.90
    3. $375
    4. $8.34
    5. $42.96
    6. $119
16. 1. $27
    2. $9.88
    3. $2600
    4. $720
    5. $32.25
    6. $299.50
17. 1. 150 feet of rope
    2. the 18-inch gold chain
    3. 8 yards of fabric
    4. the 175-foot coil
    5. ten 8-foot boards
    6. 100 feet of pipe

Daily Warm-Ups: Math in Real Life

Answer Key

18. 1. 150 square feet of denim
    2. 1100 square feet of sod
    3. 250 square feet of flooring
    4. 820 square feet of carpet
    5. paint to cover 400 square feet
    6. the 32-square-foot table
19. 1. 48 ounces of maple syrup
    2. same
    3. the 20-ounce can
    4. 1 gallon of apple juice
    5. 1 gallon of pineapple juice
    6. 64 ounces of sauce
20. 1. 2 pounds of ham for $8.50
    2. 1.5 pounds of salad for $4.50
    3. $2.29 per pound for cheese
    4. $3.79 per pound for chicken wings
    5. $6.89 per pound for salmon
    6. 3 pounds of lobster for $40.99
21. Answers will vary, but lists should include expenses for cable, food, electricity, car payments, rent, and so forth.
22. 1. 28.1% 3. 33.9%
    2. 36.2% 4. 33.9%
    5. 27.9% 6. 37.1%
23. 1. $525 4. $2281.50
    2. $104.50 5. $195.97
    3. $7187.00 6. $1614.15
24. 1. $12,730.80 4. $37,091.94
    2. $586.56 5. $16,205.08
    3. $134.58 6. $1835.88
25. 1. $5.63 4. $13.73
    2. $35.84 5. $0.69
    3. $18.55 6. $1.06
26. 1. $13,662 4. $53,840.49
    2. $713.30 5. $8817.05
    3. $769.50 6. $9908.99
27. 1. monthly interest payment = $52.08; total payment = $14,375
    2. monthly interest payment = $16; total payment = $3584
28. 1. Jon Dough 4. car repairs
    2. A-1 Motor Services 5. North Bank
    3. $421.65 6. 0072279327
29. 1. $694.70 3. 1241
    2. 12/7 4. $621.42

Daily Warm-Ups: Math in Real Life

Answer Key

30. 1. $35
    2. $285.26
    3. $13.20
    4. $493.44
    5. $3068.20
31. 1. 35%
    2. $25,078.20
    3. 33%
    4. no; higher tax rate
32. 1. 2 or 3 cents
    2. 2007: $1.02–$1.03; 2008: $1.04–$1.06; 2009: $1.06–$1.09; 2010: $1.08–$1.12
    3. 16 cents; 35 cents
33. 1. $480.00
    2. $3000.00
    3. $1200.00
34. Answers will vary, but might include age, credit history, distance traveled daily, driving record, gender, marital status, prior claims, location, vehicle make, and so forth.
35. 1. $220,000, $137,500
    2. yes
    3. yes
    4. $227,500
36. 1. enough
    2. enough
    3. $27,350 more
    4. $36,566 more
    5. enough
    6. $27,240 more
37. 1. $2.50 per square foot
    2. 3 years of service
    3. the calling card that has 750 minutes
38. 1. 576.80 yen
    2. 236.31 euros
    3. 176.91 Canadian dollars
    4. 6774.86 pounds
    5. 1116.82 Australian dollars
    6. 1863.15 Swiss francs
39. 1. $150.00 gross pay; $40.50 in taxes
    2. $93.75 gross pay; $25.31 in taxes
    3. $80.00 gross pay; $21.60 in taxes
40. 1. Store 1
    2. Store 2
    3. Store 2
    4. Store 1
    5. Store 2
    6. Store 2
41. 1. $452.16
    2. $509.25
    3. $1025.46
    4. $217.76
42. 1. $220.87
    2. $32.89
    3. $319.68
43. 1. $24,940
    2. 35%
    3. 1.98 times
    4. 35.5%, 17.1%, 35.0%, 25.0%
44. Plan 1: $2504.99
    Plan 2: $2049.99
    Plan 3: $1639.99
    Plan 4: $1274.99
    Plan 5: $529.99

Daily Warm-Ups: Math in Real Life

Answer Key

45. 1. 6% = $2098.43; 6.5% = $2212.24; 7% = $2328.56; 7.5% = $2447.25; 8% = $2568.18
    2. 6% = $755,434.80; 6.5% = $796,406.40; 7% = $838,281.60; 7.5% = $881,010; 8% = $924,544.80
    3. $82,846.80
    4. $169,110
46. 1. $1\frac{1}{3}$ cups milk, $\frac{1}{3}$ teaspoon salt, $5\frac{1}{3}$ cups coconut, 2 teaspoons vanilla
    2. $\frac{1}{3}$ cup milk, $\frac{1}{16}$ teaspoon salt, $1\frac{1}{3}$ cups coconut, $\frac{1}{2}$ teaspoon vanilla
    3. $13\frac{1}{3}$
    4. 1
    5. $2\frac{2}{3}$
    6. 64
47. 1. $2810
    2. $16,330
    3. $4012.50
    4. $8249
    5. $29,115
    6. $39,682
48. 1. $4.35
    2. $3.29
    3. $13.18
    4. $3.23
    5. $18.88
    6. $2.54
49. 1. $54.43
    2. $86.75
    3. $78.80
50. 1. 42-inch tube television
    2. 55-inch plasma television
    3. It doesn't take weight, picture quality, longevity, and so forth into consideration.
51. 24,000 BTUs is the best deal.
52. b
53. 1. 218.2 milligrams
    2. 109.1 milligrams
    3. 1181.8 milligrams
    4. 254.5 milligrams
    5. 184.1 milligrams
    6. 300 milligrams
54. 1. What kind of doctors?
    2. Do they get rewarded to do so or is it spontaneous?
    3. Is this the only banana-coconut gum they have tried?
    4. 50% less fat than what?
    5. How big is this area?
55. 1. The actual difference is only about 0.2%,
    2. Examples will vary.
56. c
57. 1. 63 miles
    2. Answers will vary. Sample answers: curves; detours; stops for gas, food, lodging

Daily Warm-Ups: Math in Real Life

Answer Key

      3. airplanes
58. 1. 150,000 toothpicks per day
      2. 83.3 kilometers per hour
      3. 54,750,000 gallons per year
      4. 14 goats per week
59. 1. 4520 feet   3. 4.3 miles
      2. 9040 feet   4. 1.1 miles
60. 1. 8.64 megapixels   4. 16.61 megapixels
      2. 3.15 megapixels   5. 5.04 megapixels
      3. 0.31 megapixels   6. 1.92 megapixels
61. 1. 2 times   3. 32 times
      2. 256 times
62. 1. 21.1 cans   3. 700 calories
      2. 41.25 pounds   4. 5 bars
63. 1. 37 calories   3. 47.3 days
      2. 74 calories
64. 1. 4 miles   3. 7.25 miles
      2. 774 calories   4. 520 calories
65. 1. 10 chocolate bars   4. 10.3 cups
      2. 7 cups   5. 720 milligrams
      3. 2.6 sodas
66. 1. 21.9 grams   3. 0.059 grams
      2. 15.6 milliliters
67. 1. 88   3. 91
      2. 68
68.

East	GB
Boston	—
New York	0.5
Toronto	5.0
Baltimore	15.5
Tampa Bay	20.0

69. 1. $750.00 per year   3. $1083.30 per year
      2. $940.00 per year
70. 1. 2^5
      2. 6^2
      3. 65^3
      4. original number of blogs × 2^7
71. 1. 10 feet   3. 31.6 feet
      2. 14.1 feet
72. 1. 4 loaves of bread, 5 packages of sandwich meat
      2. 4 packages of hot dogs, 5 packages of buns
      3. 3 packages of nuts, 2 packages of bolts
73. 1. Jackson, Leslie, Beard, Augustus, Thompson, Whitmore, Douglas, Pondexter, Taurasi, Catchings
      2. Leslie, Douglas, Whitmore, Thompson, Jackson,

Daily Warm-Ups: Math in Real Life

Answer Key

Taurasi, Pondexter, Augustus, Beard, Catchings

74. 1. 14.7 psi
    2. 44.7 psi
    3. 35.3 psi
75. 1. $0.58
    2. 6219.4 square inches
    3. $44.73
    4. 14.4 ounces
76. 1. $\frac{3}{28}$ or 0.107
    2. 0.225 pounds
    3. 5.5 times
    4. 6 whole pieces
77. 1. 16.13 mph
    2. 14.25 mph
    3. 12.59 mph
    4. 11.62 mph
78. 1. 28 mpg
    2. 9.6 mpg
    3. 66 mpg
    4. 30 mpg
    5. 18 mpg
    6. 18 mpg
79. $2781.90
80. 1. 2.8 megabytes per second
    2. 5.6 megabytes per second
    3. 9.2 megabytes per second
    4. 5.3 megabytes per second
    5. 10.6 megabytes per second
    6. 15.8 megabytes per second
81. 1. 6.7 CDs
    2. 12.1 CDs
    3. 13.4 CDs
    4. 21.3 DVDs
    5. 285.7 CDs
82. 1. $\frac{3}{4}$
    2. 0.625
    3. $\frac{1}{2}$
    4. 0.666
    5. $\frac{1}{4}$
    6. 0.157
83. 1. 129.9 square feet
    2. 9 feet, $1\frac{1}{2}$ inches
    3. $37\frac{5}{8}$ inches
84. 1. $1.00
    2. $3.36
    3. 6 cups
    4. 27.16 square feet
85. 1. $85\frac{1}{3}$ pieces
    2. 93 mini apple pies
    3. 20 hamburgers
    4. 40 sections
86. 1. 78.7 mph
    2. 89.1 mph
    3. 91.0 mph
    4. 92.0 mph
    5. 80.4 mph
    6. 85.4 mph
87. 1. $\frac{10}{1}$
    2. $\frac{100}{10}$
    3. $\frac{1000}{1}$
    4. $\frac{1,000,000}{1}$
    5. $\frac{100}{1}$
    6. $\frac{1000}{1}$
88. 1. 6:12 or 1:2
    2. 4:2 or 2:1
    3. 16:5
    4. 3:24 or 1:8,
    5. 1.5:2.5 or 3:5

Daily Warm-Ups: Math in Real Life

Answer Key

89. 1. 20; 25
    2. Whole milk has twice the calories.
    3. 2400 calories
    4. 1440 calories

90. 1.

Player	%
Mark Price	90.4
Peja Stojakovic	89.4
Steve Nash	89.6
Rick Barry	90.0
Calvin Murphy	89.2
Shaquille O'Neal	52.8
Michael Jordan	83.5

   2. Mark Price, 3. Shaquille O'Neal, 4. 6.9% lower

91. 1.

Team	Win %
New England	62.5
Miami	56.3
Buffalo	31.3
N.Y. Jets	25
Cincinnati	68.8
Pittsburgh	68.8
Baltimore	37.5
Cleveland	37.5
Indianapolis	87.5
Jacksonville	75
Tennessee	25
Houston	12.5
Denver	81.3
Kansas City	62.5
San Diego	56.3
Oakland	25

   2. Indianapolis
   3. It's a tie between the N.Y. Jets, Tennessee, and Oakland.

92. 1. 15.6% 3. 0%
    2. 31.3% 4. Answers will vary.

93. 1. 167% 3. 114%
    2. 125% 4. 123%

Daily Warm-Ups: Math in Real Life

Answer Key

94.
1. 3.98%
2. 1.99%
3. 0.16%
4. 0.8%
5. 0.32%
6. 0.06%

95.
1. 13.3%
2. 4.4%
3. 12.1%
4. 767%

96.
1. 8.3%
2. 15.2%
3. 71.4%

97.
1. $45,000, $48,000, $51,000
2. 176 mph
3. $29.28

98.
1. 2 feet
2. 10 feet
3. 3 weeks
4. 7 weeks

99.
1. 77 inches
2. 5.5 miles
3. 105 feet
4. 76,032 feet
5. 912,384 inches
6. 4 feet, 11 inches or 4.92 feet

100.
1. 16 cups
2. 2.25 pints
3. 6.25 gallons
4. 18 tablespoons
5. 5 cups

101.
1. 17.5 tons
2. 80 ounces
3. 40 tons
4. 49.5 long tons
5. 55,104 ounces

102.
1. cm or m
2. mm
3. km
4. m
5. km
6. m

103.
1. l
2. ml
3. l
4. ml
5. ml
6. l

104.
1. 1000 grams
2. 4400 pounds
3. 2273 grams
4. 0.4 grams

105.
1. 0.12 years
2. 270 minutes
3. 1533 days
4. Answers will vary; a 16-year-old would be 140,160 hours old.

106.
1. gold
2. carbon
3. iron
4. carbon
5. titanium

107. Answers will vary.
108. Answers will vary.

109.
1. Ralph Metcalfe
2. 28 years
3. 1 second
4. Eddie Tolan

110.
1. 33%
2. 29%
3. no
4. snakes

Daily Warm-Ups: Math in Real Life

Answer Key

     5. 17%
111. 1. Alvarez
     2. Brown
     3. 25%
112. Answers will vary.
113. 1. Steven
     2. Jun and Agusto
     3. no
114. 1. 71 degrees
     2. 103 degrees
     3. 6–7 P.M.
115. 1. jogging times
     2. minutes
     3. feet/minute
116. 1. $35.125
     2. $32.125
     3. $3375
117. 1. 11 A.M.
     2. 12:20 P.M.
118. 1. 6
     2. 70–79
     3. 0–19

4. 65%
5. Alvarez

4. Odina
5. Sumalee

4. 11 A.M., 7 P.M.
5. 98 degrees

4. 14–15 seconds
5. 505 feet/minute

4. $2500
5. $1800

3. It's lunchtime.
4. about 1 hour
4. 46
5. 108

119. 1.

10	00
9	55410
8	9975332
7	98664
6	651
5	8
4	5

     2. yes   4. 50s and 40s
     3. 80s   5. 81
120. 1. 6.57 meters per second
     2. 2.03 meters
     3. 7.14 meters per second
     4. 5.10 meters
121. Answers will vary.
122. 1. Answers will vary.   3. 132, 137
     2. Answers will vary.   4. 166, 171
123. 1. 28.5   3. 18.3
     2. 23.3   4. 54.2
124. 1. 68°F   4. –40°C
     2. 86°F   5. 37.7°C
     3. 212°F

Daily Warm-Ups: Math in Real Life

Answer Key

125. 1. 54 mph 3. 243 miles
     2. 5 hours, 54 minutes
126. 1. Armas: .630; Masaru: .604; Brown .601
     2. Brown has the lowest slugging average.
127. Answers will vary.
128. 1. cube, 2. hexagon, 3. sphere, 4. cone, 5. cylinder
129. Answers will vary.
130. 1–4. Drawings will vary. 5. four times larger
131. Answers will vary.
132. 1. 720 feet 3. 68 feet
     2. 22 feet 4. 246 feet (240 also ok)
133. 1. 251.2 feet 2. 20.1 centimeters
134. 1. 8100 cm² 3. 108 ft²
     2. 100 ft²
135. 1. 32 ft² 3. 4.5 million ft²
     2. 50,400 ft² 4. 89.3 fields
136. 1. 28.26 ft² 3. 12,861 cm²
     2. half the 18-inch pizza
137. Answers will vary.
138. 1. 5000 ft² 3. 225 in²
     2. 19.25 cm²
139. 1. 186 ft² 2. 2387 ft²
140. 1. 2180 in² 2. 1881 in²
141. 1. 30.4 in² 3. 14.26 ft²
     2. 24,576 µm²
142. 1. 1152 ft² 2. 412.9 ft²
143. 1. 640,000 ft²
     2. a. 705,600 ft²
        b. 544,320 ft²
        c. 161,280 ft²
144. 1. 2174 ft² 3. 248 bowling balls
     2. 21.7 in²
145. 1. 77,378.8 m² 2. 2096.9 m²
146. 1. 565.2 ft² 2. 50,554 mm²
147. 1. 381.5 in³
     2. 54 basketballs will fit in the box (25,920 in³)
     3. 8.2 in³
     4. 3161 tennis balls will fit in the box
148. 1. 163,840 ft³ 3. 120 tissues
     2. 1000 ft³
149. 1. 5.76 in³ 2. 33.5 ft³
150. 1. 144 ft³ 2. 24 more small boxcars
151. 1. 230.9 in³ 2. 42 gallons
152. 1. 2592 ft³ 2. 78,932,517 ft³
153. 1. 60.3 feet 2. 5 feet
154. Answers and graphs will vary.

Daily Warm-Ups: Math in Real Life

Answer Key

155. 1. to know what to play to connect advertisers with listeners
 2. pop/rock
 3. not necessarily a random group
 4. young people or teenagers
156. Answers will vary.
157. Answers will vary.
158. 1. 1:345   3. 1:2760
 2. 1:8
159. 1. 15:374 or 4%
 2. 1:548,354,040 or 0.00000018%
 3. 138:15,283 or 0.9%
160. 1. Class 1: 82.4; Class 2: 77.4
 2. yes
161. 1. 120   2. 2184
162. 1. 5   4. 2
 2. 4   5. 1
 3. 3   6. 120
163. 1. 162 kinds   2. 168,000 meals
164. 1. 20   2. 792
165. 1. 28   3. 35
 2. 6   4. 1:8
166. 1. 210 passwords
 2. 252 passwords
 3. Answers will vary but should refer to increasing the number T and/or the number S.
167. 1. many; 24 combinations
 2. short; 120 combinations
 3. jumble; 720 combinations
 4. section; 5040 combinations
 5. dreaming; 40,320 combinations
 6. badmouths; 362,880 combinations
168. 1–4. Answers will vary; e, t, and a should be the three most common letters found.
 5. Letters appear with different frequencies and are grouped accordingly.
169. 1–2. Answers will vary.
 3. 125 pounds
170. 1–2. Answers will vary.
 3. 30 inches
171. 6-foot cords
172. 1. Emilio: 15.6; Doug: 13.1
 2. Emilio: 2.8; Doug: 2.6
 3. Emilio: 1.5; Doug: 1.4
 4. Emilio
173. 1. 1:8

Daily Warm-Ups: Math in Real Life

Answer Key

2. yes

3. No; for every 7 times you get a quarter, he should only get 4 back.

174. 1. 1:6 3. 1:7776
     2. 1:216

175. 1.

Die value	1	2	3	4	5	6
1	2	3	4	5	6	7
2	3	4	5	6	7	8
3	4	5	6	7	8	9
4	5	6	7	8	9	10

2. 2–1:24; 3–1:12; 4–1:8; 5–1:6; 6–1:6; 7–1:6; 8–1:8; 9–1:12; 10–1:24

176. 1. 1 in 1.46 billion 3. 1 in 1.30 billion
     2. 1 in 519 million

177. 1. 2778 women 3. 27,370 men
     2. 14,698 women

178. 1. 1:2; 1:24
     2. 1:7,330,554,000; 1:73,305,540

179. 1. 1:13 2. 1:221

180. .3965694566

Daily Warm-Ups: Math in Real Life